ACCOUNTABLE

ACCOUNTABLE

A CFO's Framework for Governing AI Before You're Asked to Defend It

Brian H. Zahn

Pacific Meridian Press
2026

Published by Pacific Meridian Press
Ventura County, California

The views expressed in this work are solely those of the author and do not represent the views or opinions of any organization, employer, or affiliated institution.

ISBN (eBook): 979-8-9955208-0-1

First Edition | June 2026
Printed in the United States of America

To my fellow executives...

the CFOs, Controllers, CIOs, CEOs, and board members navigating an era no one fully prepared us for.

AI did not wait for permission to enter our organizations. It arrived quietly, through a subscription line item, an embedded feature, a developer's shortcut. And suddenly it was everywhere, consequential, uncharted, and ours to account for.

This book is dedicated to every leader who has decided that "I don't fully understand it yet" is not a reason to defer governance, but a reason to begin it.

The organizations that come through this era with their integrity intact will be led by people like you.

Table of Contents

Author's Note

There was no single courtroom moment.

No deposition. No regulatory summons. No board meeting that went catastrophically wrong. What pushed me to write this book was quieter than any of those things, and I suspect more familiar to the executives who will read it.

There was a gap. And it kept getting wider.

I watched AI move from a technology conversation to an operational reality in my organization, in peer organizations, and across every industry I have worked in over my career. And in every room, the same pattern played out. Sophisticated executives. Serious people. Genuine commitment to their fiduciary obligations. And no framework for translating those obligations into the specific demands that AI now creates.

I looked for a book that addressed that problem directly, written for CFOs, Controllers, CIOs, CEOs, and board members, not for technologists or compliance specialists. A book that started from governance, not from code. I could not find it.

So I wrote it.

What follows is not a technology manual. It is a governance framework built on fiduciary principles you already operate by, extended into the obligations that AI now places on every executive who oversees it. You do not need to understand how these systems are built to govern them. You need to know what questions to ask, what accountability structures must exist, and what your board needs to see.

By the time you finish this book, you will have all of that.

The gap I kept watching grow is now closing, not because organizations got more careful, but because regulators, courts, and boards are beginning to demand the accountability structures that should have been built already.

You are reading this book. That tells me you have already made the more important decision.

Let us get to work.

Brian H. Zahn

Chief Financial Officer

Ventura County, California

INTRODUCTION

The Signature Problem

Why Being Smart Is No Longer Enough

By the time you close this book, you will be able to do three things that most senior executives in your industry currently cannot.

1. Walk into a board meeting and answer, without hesitation, exactly what AI systems are running in your organization, who is accountable for each one, and what your governance posture looks like under regulatory scrutiny.

2. Demonstrate, with documentation, that you have exercised fiduciary oversight over AI, the standard courts and regulators have already established, not the standard everyone is still waiting to be formally told about.

3. Build a governance program that outlasts you: one embedded in policy, role, and culture rather than held together by your personal commitment to it.

None of those capabilities require technical expertise. They require a governance structure, the kind this book is built to deliver. The story below shows you exactly what its absence costs.

Nobody told her she was responsible for it.

> **Illustrative Scenario:** *The following scenario is representative of documented patterns in executive AI governance. Names, organizations, and specific details are illustrative.*

That was the part that kept surfacing in the deposition, not the damage, not the forty-one thousand customers whose loan applications had been processed by a credit-scoring supplement the vendor had marketed as "decision support." A tool, as the plaintiff's attorney noted, that had been supporting decisions for three years without a single human at her company ever reviewing its outputs. The part that kept surfacing was that nobody had told her.

The court did not see it that way. Neither did her board.

She had been CFO for six years. By every conventional measure, she was excellent at the job, rigorous on reporting, trusted by analysts, respected by her team. Her signature was on the vendor contract. She had approved the budget three years running. The technology team managed the system. She governed everything else. When the complaint named her personally, she told outside counsel the seven words that ended her tenure:

"I had relied on our technology team."

> ***"I had relied on our technology team."***

The judge's response was not angry. It was almost clinical. She said: "CFOs do not delegate fiduciary responsibility. They exercise it."

This book begins there. Not because I want to frighten you, and not because your situation is identical to hers. It begins there because that sentence, CFOs do not delegate fiduciary responsibility, they exercise it, is the sentence the entire AI governance conversation has been circling for three years without landing on directly. You are already living in the legal and organizational environment that sentence describes. The only question is whether you are equipped for it.

That is the signature problem. Not that she signed the contract. That she believed her signature transferred the governance.

I wrote this book because most executives face the same gap, not because they are careless or disengaged, but because the accountability standard moved faster than the governance norms, and no one gave them the architecture to catch up.

THE EXACT PROBLEM

Here is what I hear from executives, in some version, in almost every serious conversation about AI governance:

"I know this matters. I'm not sure I know what to do about it."

That is not a sentence spoken by people who are disengaged or careless. It is spoken by people who have been running hard in a role that has genuinely changed under their feet. These are not people who avoid hard problems. They are people who have not yet been given the right frame for this one.

That is the gap this book was written to close. And I want to name it with precision, because how you name a problem determines how you solve it.

The problem is not that executives do not understand AI. Technical fluency in your technology team does not produce governance accountability in your organization. The problem is not that executives do not take AI seriously. The investment levels, the board conversations, the hiring, the headlines, AI has been on every agenda. The problem is that taking something seriously and having a governance structure for it are not the same thing. For most organizations, the governance structure does not exist at the level the fiduciary obligation requires.

What does that mean in practice? It means that if your board chair asked you, today, to describe your organization's AI governance posture, not your AI strategy, not your AI investment, but your governance posture: what systems are running, who owns each one, how risk is tiered, what the total liability exposure looks like, and what accountability structure surrounds it, most executives could not produce a complete, defensible answer. Not because they have not been paying attention. Because the structure that generates that answer has not been built.

> ***Taking something seriously and having a governance structure for it are not the same thing.***

The cost of that gap has been abstract long enough to defer. That is changing. Regulatory frameworks are moving from advisory to mandatory. Litigation is producing precedent. Board expectations are sharpening. The executive treating AI governance as a technology department responsibility is operating on borrowed time, not in a theoretical future, but in the legal and organizational environment that already exists today. Deferral is no longer a low-cost option. It is a compounding liability.

THE ROOM

I want to tell you about a specific moment that changed how I think about this problem, because it is more instructive than any statistic.

I was invited to speak at an executive leadership forum, roughly two hundred senior leaders across industries, the kind of room where the collective organizational responsibility probably exceeded a trillion dollars. Before I started, the moderator, a former regulator who now advises large institutions on governance, asked the room a single question. She did not ask it to embarrass anyone. She asked it because she already knew the answer, and she wanted the room to know it too.

She said: "How many people in this room, right now, without looking anything up, can tell me: what AI systems are running in your organization, who is the named accountability owner for each one, and what data each system touches?"

Four hands went up. Out of roughly two hundred.

I watched the others. They did not look embarrassed. They looked like people who had just noticed a blind spot, the sudden, uncomfortable awareness of something they had been looking past so long that not-seeing it had become a habit.

After the session, I spoke with two of the four who had raised their hands. I wanted to understand what they had done differently, because they were not technically more sophisticated than the rest of the room, both said so immediately. What they had in common was not technical expertise. One had built a formal AI inventory eighteen months earlier after a regulatory examination flagged the gap. The other had gone through a vendor incident that required her to document every AI system in her organization as part of a remediation response. One of them told me the examination had cost her organization fourteen weeks and two outside counsel firms. She said she would have paid twice the cost to have started before the examination began. Neither raised their hand because they understood machine learning. They raised it because an external forcing function had required them to build a governance structure, and the structure had given them the language.

This book is the forcing function without the incident.

The other hundred-and-ninety-six executives were not negligent. They were operating the way most organizations operate, AI in the hands of the technology function, governance assumed rather than documented, oversight implicit rather than structured. The accountability assumption runs like an underground current through most executive teams: someone is taking care of this. That assumption may be the most expensive belief in modern corporate governance. Because in the AI context, "someone is taking care of this" almost never means what the executive in the corner office thinks it means. The technology team is managing the technology. No one is exercising the fiduciary oversight. And when something goes wrong, and in a portfolio of AI systems operating at scale, something will, the person whose name is on the contract does not get to cite the assumption.

THE REFRAME

Everything in this book rests on a single idea. I want to give it to you before a single framework has been introduced, because it changes how you read every chapter that follows.

> ***THE CORE PREMISE***
>
> ***AI governance is not a technology problem. It is a fiduciary problem. And fiduciary problems are ones you already know how to solve.***

This reframe matters because of what it frees you from. If AI governance is a technology problem, you need technical expertise to address it. You are dependent on the people who have it, always one layer removed from the real substance, and the best you can do is ask good questions and hope the answers are complete. If AI governance is a fiduciary problem, you bring the same tools, the same discipline, and the same accountability framework you apply to every other domain where your organization bears material risk.

You have governed capital allocation without personally underwriting every transaction. You have managed regulatory exposure without interpreting every statute. You have overseen legal risk without litigating every case. You did it by building structures, assigning accountability, commissioning expertise, and holding the outcomes. AI governance works exactly the same way. The technical layer is real and it matters, but it is not your layer. Your layer is the governance layer. And the governance layer does not require you to understand model training or system architecture. It requires you to know what the system is deciding, who is accountable when it decides wrong, what the exposure looks like, and whether your board can see that responsible oversight exists.

The framework this book builds is structured around eight governance obligations. The acronym is FIDUCIAR, and each letter names a fiduciary act that an executive who governs AI responsibly must perform. You will work through each one in the chapters ahead, but here is the architecture so you can see where you are going:

> ***F. Frame as Fiduciary Obligation.*** *AI governance is a fiduciary question first, and a technology question second. Frame it at the level of governance, not engineering, every time it arises.*
> ***I. Inventory All AI Systems.*** *Every AI system in your organization, authorized or not, documented in a single governance record with named accountability owners.*
> ***D. Diagnose Risk Posture.*** *Not uniformly. AI systems are not equally risky, and governing them as if they are wastes resources and misses the actual exposure. A tiered risk model allocates governance attention where it belongs.*
> ***U. Understand Total Cost of Ownership.*** *The subscription is the smallest number. The full cost includes infrastructure, talent, compliance, liability reserve, and the organizational attention that never appears on a P&L.*

> ***C. Create the Accountability Chain.*** *Accountability must run from the vendor contract to the board report with a named human at every link. When it does not, you have policy intent. When it does, you have governance.*
> ***I. Implement the 90-Day Audit.*** *The governance program begins with a defined ninety-day sprint. The sprint produces the foundation, not the finish line, and it puts a date on the calendar for when governance becomes real.*
> ***A. Audit Ongoing.*** *Override rates, vendor performance, output sampling. Governance is real when the audit is recurring, not when the policy is filed.*
> ***R. Report to the Board in Governance Language.*** *A board-ready AI governance summary is a document any director can read, interrogate, and rely on. Most organizations do not have one.*

Eight obligations. The architecture of everything that follows.

I want to be explicit about what this book will not do, because I have seen the misunderstanding cause real damage. This book will not help you evaluate AI vendors on technical merit, decide whether to build or buy, or assess model performance characteristics. There are good resources for those questions. They are not this book. This book is for the executive who needs to govern AI, who needs to demonstrate, to a board, a regulator, or a court, that responsible oversight exists. Technical competence in your technology team is not a substitute for fiduciary accountability at the executive level. This book builds the latter.

BEFORE YOU REACH CHAPTER ONE

Most books ask nothing of you until the end. This one makes a different request: arrive at Chapter One with a clear-eyed picture of where your organization actually stands. Not a polished one. An honest one. Four things you can do before tomorrow.

1. **Answer the room question.**

Without looking anything up, write down every AI system you can name in your organization. Who owns each one. What it decides. What data it touches. Do not ask your technology team. Do not open a folder. Write what you personally know, right now.

The most common mistake: deciding this list is someone else's job to build. You can delegate the research. You cannot delegate the accountability. What you write, or cannot write, in the next five minutes is your personal governance baseline. Every blank is a gap you will close. Success looks like honesty: a short list, possibly very short, and the clear-eyed recognition that Chapter Four is going to matter.

2. **Locate the last AI-related document that came before your board.**

Find it. Read it in sixty seconds. Ask one question: does this document demonstrate governance, or investment? Governance addresses risk, accountability, and oversight. Investment addresses cost, capability, and ROI. Both have value. Only one satisfies the fiduciary standard.

The most common mistake: conflating the two. A compelling AI strategy presentation is not a governance document. A vendor ROI analysis is not an accountability structure. If what you find is investment documentation, that is useful information. It tells you the exact distance between where you are and where this book takes you.

3. **Name one person.**

Who in your organization would receive the call, at any hour, if an AI system caused a significant failure tonight? Not the vendor. Not the technology team generically. One person, by name, whose role explicitly includes being the fiduciary accountability owner for AI outcomes. If you cannot name that person in thirty seconds, your accountability chain does not yet exist.

The most common mistake: naming your CTO or CIO by reflex. They are likely the right technical escalation point, that is, different from being the governance accountability owner. The distinction matters, it is not obvious, and Chapter Seven will make it exact. For now, notice whether you have an answer at all.

4. **Make one decision.**

Decide that you are going to finish this book. Not skim it, not delegate it, not add it to the stack. Finish it. The executives in that room who could raise their hands did not have more hours than you. They had a forcing function that made the work non-negotiable. Let this be yours.

The most common mistake: treating this as one more resource to consume rather than a structure to build. Every chapter is a construction project. The chapters do not produce ideas. They produce documents, tools, and records. When you finish, something will exist in your organization that did not exist before. That is the commitment this book asks for.

WHAT THIS BOOK PROMISES

The CFO who found out in discovery was not careless. She was unstructured. The hundred-and-ninety-six executives who could not raise their hands were not disengaged. They were ungoverned in a domain that now demands governance. The gap between understanding that AI creates liability and being able to demonstrate, with specificity, that you have exercised oversight, that is, the gap every chapter in this book is designed to close.

Here is the single most important insight I can offer you before you turn the page: the difference between the executives who raised their hands and those who did not was not knowledge. It was structure. Structure is buildable. You are holding the blueprint.

By the time you close this book, you will be able to walk into a board meeting and answer the room question without hesitation, not because you have become a technologist, but because you have done what accountable executives do. You have built the structure, documented the oversight, and owned the outcome.

Chapter One opens with the legal landscape, what courts and regulators have already decided about executive accountability for AI. Before you have built a single governance structure. Before you have reviewed a single policy. I start there deliberately, because the fastest way to understand what this work is worth is to understand what its absence has already cost people who were, by every prior measure, excellent at their jobs.

> ***You are not behind because you have been inattentive. You are behind because the accountability standard moved faster than the governance norms. This book moves you forward.***

The lawsuits are already in the federal docket. The board questions are already being asked. The regulatory frameworks are already in motion. The executive who has done this work answers those questions with documented, defensible specificity. The executive who has not answers them the way she did, with seven words that were the most expensive of her career.

The governance program you build in the next seventeen chapters will be built once. Make it the version that holds under examination, not the version that was assembled quickly because the examination was already scheduled. The difference between those two programs is not the governance tools. It is the intention with which you start.

This book is not a resource to consume. It is a structure to build.

BOARDROOM TAKEAWAY | *The Last Executive Standing*

What a board member should be able to ask:

- What documentation exists today that would allow us to demonstrate fiduciary oversight over our AI systems to a regulator or court?
- Which of our AI deployments creates the greatest concentration of liability under the four established categories?

What a CFO or risk leader should have documented:

- A written mapping of your organization against the four AI liability categories
- The identity of the highest-risk AI deployment, by name, with a single sentence on its current governance status

What evidence should exist:

- A scheduled governance conversation with your CRO or General Counsel within the next two weeks
- A document showing whether your board's most recent AI briefing addressed governance or capability

CHAPTER ONE: PART I: THE RECKONING

The Last Executive Standing

What the Lawsuits Already Tell Us

The legal standard governing executive accountability for AI has already been established. Not proposed. Not pending. Established, in courts and regulatory guidance that most senior executives have not read, because the cases have not yet been widely covered in the business press. That gap, between what the law requires and what most executives believe is expected of them, is the subject of this chapter.

Here is what that standard looks like when it is applied.

> **Illustrative Scenario:** *The following scenario is representative of documented patterns in executive AI governance. Names, organizations, and specific details are illustrative.*

The deposition lasted eleven hours.

The executive on the other side of the table was not a reckless person. He had spent twenty-three years building a career that, by any reasonable measure, reflected the qualities you would want in a senior leader: deliberate, thorough, respected by peers and subordinates alike. He had overseen regulatory examinations, navigated two recessions, and managed a public company through a proxy fight. He was not someone who cut corners. He was not someone who ignored risk.

He was someone who had never been asked, in any formal context, to account for the AI systems running inside his organization. Nobody had built the structure that would have required him to answer that question. Nobody had ever handed him a document that said: here is what we are running, here is what it is deciding, here is who is accountable when it is wrong. He had assumed those things were managed. They were not. And for eleven hours, in a conference room that smelled of recycled air and bad coffee, a plaintiff's attorney asked him to explain decisions that a model had made, autonomously, about real people's financial futures.

He had no good answers. Not because he was unintelligent. Because no one had built him the structure to produce them.

This chapter is about what happens when that structure is absent. Not theoretically. Not eventually. Now. Because the legal and regulatory environment that this executive found himself inside already exists. The cases are already in the docket. The precedents are already being set. And the standard of care that courts and regulators are applying to senior executives in AI-related matters has moved in one decisive direction: from what did you know to what should you have known. That shift changes everything. It changed everything for him. It will change everything for you, unless you act before it applies.

WHY EXECUTIVES ARE STILL UNPREPARED

The reason most executives are not adequately prepared for this legal landscape is not, as I said in the introduction, inattention or incompetence. It is something more structural: the AI governance conversation, in most organizations, has been owned by the technology function, and the technology function does not think in fiduciary terms. It thinks in operational terms. Performance metrics, uptime, model accuracy, error rates. These are genuinely important. They are also not the questions a court asks.

A court asks: who was responsible for knowing what this system was doing? Who had the authority to stop it? What governance structure existed to catch the problem before it became a harm? What records exist to demonstrate that the organization exercised oversight? When the technology team says "the model performed within acceptable parameters," and the plaintiff's team says "the model denied loans to qualified applicants at twice the rate in one demographic group," those are not contradictory claims. Both can be true at once. And the question of who is accountable for that outcome lands not on the model, and not on the technology team, but on the executive who had the organizational authority to know and the fiduciary obligation to govern.

Most executives have not sat with that reality long enough to feel its full weight. In part because the cases are not widely reported in the business press the way financial fraud or product liability cases are. In part because AI-related litigation is still, in historical terms, early, the volume of cases is accelerating but the household names have not yet multiplied to the point where they appear on the front page of publications that senior executives read. And in part because there is a comfortable story available: my company is not using AI in that way, we are not in that industry, our technology team has this handled.

> ***The standard of care has shifted from 'did you know?' to 'should you have known?' That shift is not coming. It is here.***

"I trust them" is not a governance answer. It is exposure with a comfortable feeling attached. This chapter is designed to remove that feeling and replace it with something more durable: an accurate understanding of the legal terrain, and the beginning of the structure that lets you walk across it confidently.

FOUR CATEGORIES OF LIABILITY. ONE STANDARD OF CARE.

The legal landscape for AI-related executive accountability is not an undifferentiated cloud of risk. It has a shape. It has established categories. It has documented cases showing exactly how courts and regulators think about who is responsible when AI causes harm. Understanding those categories is not a legal exercise. It is a governance exercise, because each category maps directly to a governance obligation that a reasonable executive, exercising ordinary care, should have in place.

Four categories of AI-related executive liability are already established in US and EU precedent. Each is defined plainly below. None requires you to understand how a model works. All require you to understand what your organization has built, deployed, and allowed to operate.

> ***CATEGORY 1, DISCRIMINATORY OUTPUTS***
>
> *When an AI system produces decisions or recommendations that disadvantage people on the basis of race, gender, age, national origin, or other protected characteristics, the organization, and potentially its senior executives, bears liability for those outcomes. The fact that a model produced the discrimination, rather than a human being, is not a defense. Courts and regulators have consistently held that*

organizations are responsible for the outputs of systems they deploy, regardless of how those outputs were generated. Executive liability attaches when it can be shown that governance structures capable of detecting the discriminatory pattern either did not exist or were not functioning.

PRECEDENT

Multiple US financial institutions have faced enforcement actions from federal regulators, including the CFPB and DOJ, for AI-driven lending decisions that produced statistically significant disparate outcomes by race and national origin. Settlements have reached nine figures. Consent orders have named executives by role where oversight documentation was absent.

CATEGORY 2, DATA PRIVACY VIOLATIONS

AI systems consume data, often extraordinary amounts of it, from sources not always gathered with AI processing in mind. When an AI system processes personal data in ways that violate applicable privacy law, whether GDPR in Europe, CCPA in California, HIPAA in healthcare, or sector-specific federal requirements, liability attaches to the organization regardless of whether the violation was intentional. Executive exposure increases when records show that the data governance structure was inadequate relative to the known processing activity: when the AI system was ingesting protected categories of data, and no one in a governance role had evaluated whether the legal basis for that processing existed.

PRECEDENT

EU regulators have issued GDPR enforcement actions against organizations whose AI systems processed health and behavioral data without adequate legal basis, with fines calculated as a percentage of global annual revenue. US state regulators have cited AI-related data processing violations in enforcement actions under CCPA and related statutes. In each case, the absence of documented executive-level oversight of data processing decisions was a factor in both the finding and the penalty.

CATEGORY 3, NEGLIGENT DEPLOYMENT

An AI system deployed without adequate testing, without a defined use boundary, or without documented risk assessment constitutes negligent deployment when it causes harm. The negligence does not require intent, only that the deployment decision was made without the care a reasonable organization, exercising ordinary fiduciary diligence, would have applied. This category is particularly relevant for AI systems operating in high-stakes domains: credit decisions, hiring, medical triage, insurance underwriting, pricing, or any context where the system's output materially affects a person's access to opportunity or resources. Negligent deployment cases focus on the organizational decision to deploy, which is an executive decision, not a technology decision.

PRECEDENT

State court litigation in the US has produced early findings against employers whose AI hiring tools screened out candidates in patterns correlated with protected characteristics, where the employer could not demonstrate pre-deployment fairness testing. EU AI Act enforcement, now active, establishes affirmative deployment documentation requirements for high-risk AI applications, creating a clear evidentiary standard for negligent deployment claims.

CATEGORY 4, FAILURE OF OVERSIGHT

This is the category that will define AI-related executive accountability for the next decade, and the one least discussed in boardrooms today. Failure of oversight attaches when an AI system operates within an organization without adequate human review of its decisions, without a monitoring mechanism capable of detecting systematic errors or harmful patterns, and without a governance structure documenting who is responsible for those functions. Unlike the first three categories, failure of oversight does not require a specific harmful output as its trigger. It attaches to the governance gap itself. It is the corporate law equivalent of the judge's statement from the introduction: fiduciaries do not delegate responsibility, they exercise it. The question is not whether something went wrong. The question is whether the structure existed that would have caught it if it had.

PRECEDENT

SEC guidance on AI disclosures for public companies, FDIC and OCC supervisory letters on model risk management for financial institutions, and EU AI Act Articles 9 and 17 all establish affirmative oversight obligations for organizations deploying AI in covered domains. Multiple enforcement actions have cited the absence of documented human oversight mechanisms as a primary basis for findings. Courts in negligence cases have begun applying a 'should have known' standard to AI system behavior that falls within the range of foreseeable outcomes for the application.

Read across those four categories and a single governance requirement emerges: the executive accountable for an AI system must be able to demonstrate, with documentation, that the system's outputs were known, its risks were assessed, its deployment was deliberate, and its ongoing operation was under meaningful oversight. Not the technology team's oversight. Executive oversight. Documented oversight. The kind that produces a record.

That is exactly the standard you already apply to other material risks in your organization, the same discipline you bring to credit risk, litigation risk, and regulatory compliance, applied to a domain where most organizations have not yet built the equivalent structure. The chapters that follow build it. This chapter establishes why it is non-negotiable.

THE RISK COMMITTEE CHAIR

I want to tell you about an executive who got this right, not because she was more technically sophisticated than her peers, but because she asked one question nobody else in her organization had thought to ask.

> **Illustrative Scenario:** *The following scenario is representative of documented patterns in executive AI governance. Names, organizations, and specific details are illustrative.*

She was the chair of the risk committee at a mid-size financial institution. The organization had been deploying AI-driven tools for several years: a credit scoring supplement, a fraud detection model, a customer service routing system, and, more recently, a collections optimization tool that determined the timing and channel of outreach to customers in arrears. The technology team managed all four. They reported to the CTO. The CTO reported, periodically, to the board on technology strategy. AI came up in those reports as a capability story. Investment. Efficiency gains. Competitive positioning.

At a risk committee meeting in early 2023, she interrupted the agenda. Not dramatically. She simply asked: "Do we have a document that lists every AI system we operate, what each one decides, and who is accountable when one of those decisions causes harm?"

There was a silence of the type I described in the introduction. The CTO said he believed that documentation existed at the model level. The Chief Risk Officer said he could pull together a summary. The General Counsel said she would want to review whatever was produced before it went to the committee.

Six weeks later, the committee received a document. It listed nine AI systems. The four the risk committee chair knew about, and five more that had been deployed by business lines without formal technology review, including a pricing optimization tool in the consumer lending division that had been running for fourteen months and whose outputs had never been reviewed for fair lending compliance.

The risk committee chair called an emergency session. She did not call a press conference. She did not escalate to the board before she understood what she was looking at. She commissioned a targeted compliance review of the pricing tool's historical outputs. It took four weeks. The review found a statistically significant pattern in pricing decisions that correlated with geography in ways that tracked, imperfectly but meaningfully, with the demographic composition of the applicant pool. The pattern was not intentional. It was a product of the training data. It was also a fair lending problem.

The organization self-reported to its primary regulator. Before the examination team arrived, the risk committee chair was able to hand them a document showing: the system, when it was identified, the governance gap that had allowed it to operate without oversight, the compliance review that had been commissioned, the findings, and the remediation plan. The examination team noted, in their report, that the organization's response demonstrated "appropriate governance awareness and proactive remediation consistent with the expectations of a supervised institution." No enforcement action was taken.

Now consider what happens if she does not ask that question. The pricing tool continues to run. The pattern continues to compound. A regulatory examination discovers it eighteen months later, in the context of a broader fair lending review. The documentation shows no internal awareness, no monitoring, no oversight. The enforcement action that follows is not about the pattern in the pricing tool. It is about the failure of oversight. It is about the governance gap. And the names in the consent order are not the technology team's.

One governance question. One governance structure. One record that demonstrated oversight. That is fiduciary work. That is exactly what this book builds.

WHAT YOU CAN DO BEFORE TOMORROW

Understanding the legal landscape is necessary. It is not sufficient. The executive who reads this chapter and returns to their inbox without taking a single concrete action has added anxiety without adding protection. Four actions that move you from informed to positioned.

1. **Pull the last AI-related document your board or risk committee received.**

Find it. Read it through the lens of the four liability categories above. Does it address discriminatory outputs? Does it establish data governance relative to AI processing? Does it document the basis for deployment decisions? Does it name the oversight structure? If the document does none of those things, if it is a capability brief, a budget summary, or an ROI analysis, you now know exactly what is missing. The gap between that document and what the four liability categories require is the governance gap this book closes.

Governance documents are not produced by the technology function acting alone. They require a mandate. That mandate comes from you. The absence of a governance document is the governance problem you are now authorized to solve. Success looks like knowing the gap before you name it: a short stack of documents, each one revealing exactly how far the governance program needs to travel.

2. **Map your organization against the four categories.**

For each of the four liability categories, discriminatory outputs, data privacy, negligent deployment, failure of oversight, write a single sentence answering this question: if a regulator or plaintiff's attorney asked me to produce documentation demonstrating that our organization has addressed this category, what would I hand them? If the answer is "I would have to ask my technology team," that is not documentation. That is a delegation. You need the document.

Most common mistake: conflating a policy with a governance record. A responsible AI policy that says 'we are committed to fairness and transparency' does not address Category 1. An audit log showing that the outputs of a high-stakes AI system were reviewed for demographic parity on a quarterly basis does. The policy is the intention. The record is the governance. Courts care about the record.

3. **Identify your highest-risk AI deployment.**

Using the four categories as a lens, identify the one AI system in your organization that creates the greatest concentration of liability. Not the most expensive, not the most visible, not the most recently deployed. The one whose failure, discriminatory output, privacy violation, negligent deployment finding, or oversight gap, would produce the most significant consequences for the organization and for you personally. Name it. Write it down. Chapter Five builds the full risk tiering model, but this single identification is worth doing today because it focuses your attention on the right problem.

Risk concentration tracks decision autonomy and consequence magnitude, not visibility. A customer-facing chatbot may be more prominent. An internal underwriting model that processes loan applications autonomously is almost always more legally consequential. Name the most dangerous system, not the most visible one. Success looks like a single named system whose governance gaps, once closed, would produce the most meaningful reduction in your personal fiduciary exposure.

4. **Schedule a governance conversation, not a technology briefing.**

In the next two weeks, schedule thirty minutes with your CRO, General Counsel, or equivalent, not your CTO. The agenda is a single question: what documentation exists, today, that would allow us to demonstrate to a regulator or court that we have exercised fiduciary oversight over our AI systems? You are not asking for a technology assessment. You are asking for a governance inventory. The distinction will be immediately legible to whoever you ask, and the answer will tell you exactly how much work the chapters ahead need to do.

Call it a governance conversation, not a strategy conversation. Be explicit about the distinction when you schedule it. Strategy conversations produce roadmaps. Governance conversations produce records. You need a record. If you arrive without naming that distinction, you will receive exactly the capability brief you already have. Success looks like a thirty-minute conversation that ends with a written list of what governance documentation exists and what does not.

THE STANDARD HAS ALREADY MOVED

The executive in that deposition room was not unlucky. He was ungoverned. The cases that established the four liability categories were not aberrations. They were indicators, of a legal environment that has already decided, before your organization's specific situation ever comes up, that senior executives are responsible for the governance of AI systems operating under their authority, regardless of whether anyone explicitly told them so.

The single most important insight in this chapter is this: ignorance of what AI is doing inside your organization is no longer a legal defense. It is evidence of a governance failure. The standard of care has moved. The question is not whether you will be held to it. The question is whether you will be prepared when it is applied.

> ***The executive who has done this work answers the hard question with a document. The executive who has not answers it with eleven hours in a deposition room.***

You now know what the legal landscape looks like. You know the four categories of liability that are already established, the direction the standard of care is moving, and the governance record that distinguishes the executive who acted from the one who testified.

What you do not yet have is a complete picture of what you are actually governing. The legal exposure you just mapped attaches to AI systems, and most organizations are running more of them than their senior executives know about. Some were formally approved. Some were deployed by business lines without a technology review. Some arrived embedded inside software your organization was already using, without a single deliberate deployment decision being made. Chapter Two is about those systems: what they are, where they come from, and why the ones you have not found yet carry exactly the same legal exposure as the ones you have.

The map is not a reassuring document. Most organizations that complete it find more than they expected, systems deployed without review, tools embedded in software renewals, models built by developers who have since left, and a governance record that does not match the actual operational footprint. Chapter Two is about finding all of it, before someone else does.

BOARDROOM TAKEAWAY | *Shadow AI*

What a board member should be able to ask:

- Do we have a complete inventory of every AI system operating in our organization, including those that arrived without formal review?
- What AI capabilities have been enabled in our largest SaaS platforms in the past twenty-four months?

What a CFO or risk leader should have documented:

- Results of a corporate card spend analysis filtered for unapproved AI subscriptions
- A preliminary Shadow AI list from the five detection methods: spend analysis, vendor update review, developer survey, contract review, and API audit

What evidence should exist:

- A briefing record with General Counsel confirming that a Shadow AI detection exercise has been initiated
- At least one Shadow AI system that has been identified, named, and added to a preliminary governance record

CHAPTER TWO: PART I: THE RECKONING

Shadow AI

What's Running in Your Organization Right Now

Your marketing team is using an AI tool that ingests customer data. Right now. Today.

You did not approve it. Your IT department did not review it. Your legal team did not evaluate whether it is compliant with your data processing agreements, your privacy policy, or the regulatory framework that governs your industry. One of your marketing managers discovered it six months ago, signed up with her corporate card, invited six colleagues, and has been running customer behavioral data through it ever since because it produces better campaign segmentation than anything your organization has officially licensed. She is not malicious. She is not reckless. She found a tool that works, used the purchasing authority she already had, and solved a problem her team had been asking the technology department to solve for two years.

She has no idea she created a liability. Neither do you. That is the problem this chapter was written to solve.

From her seat one level down, the same tool raises a different question, not the organization's exposure, but her own value in an AI augmented workplace. That question is the subject of *Indispensable*, the companion volume to this book.

The conversation about AI governance almost always begins with the AI an organization knowingly deployed. The model the board approved. The vendor contract that went through procurement. The use case that got a formal technology review. Those systems matter, and the chapters that follow will give you the tools to govern them. But the more urgent problem, for most organizations, is not the AI they know about. It is the AI they do not. The systems that arrived without a ticket, without a review, without anyone in a governance role ever asking: what is this, what data does it touch, and what is it deciding? That population of systems. Shadow AI, is larger in most organizations than the officially sanctioned AI portfolio. And it carries every one of the four liability categories from Chapter One, without a single piece of governance documentation to show for it.

THE GOVERNANCE VACUUM NOBODY NAMED

There is a reason Shadow AI has proliferated so rapidly, and it is not that your employees are cavalier about risk. It is that the organizational conditions that produce Shadow AI are the same conditions that produce every other form of ungoverned behavior in large organizations: capability moved faster than policy, and nobody with governance authority noticed in time to build the guardrails before the behavior became embedded.

Consider the timeline. Three years ago, most of the AI tools that now exist as consumer-facing, credit-card-purchasable, no-IT-review-required applications did not exist at all. The tools that did exist were largely the province of data science teams, technical enough that only technical people were using them, which meant the governance gap, to the extent it existed, was at least bounded by a capability barrier. You had to know how to build a model to run one. That barrier is gone. The tools available today to a marketing manager, a finance analyst, a recruiting coordinator, or an operations supervisor with no technical background whatsoever are more capable than the enterprise AI systems most organizations were running at considerable cost five years ago. And they are available for the price of a monthly subscription that sits well below the threshold that triggers a procurement review.

> ***Shadow AI is not a technology problem. It is a governance vacuum. And a vacuum, by definition, is what exists when no structure has been built to fill it.***

The cost of that vacuum is not theoretical. It is actuarial. Every AI system operating in your organization without governance documentation is a liability that has not been inventoried, a risk that has not been tiered, a data processing activity that has not been reviewed, and an accountability gap that has no named owner. When something goes wrong, a data breach, a discriminatory output, a vendor whose model was retrained without your knowledge and whose new behavior introduced a compliance problem, the question of who is responsible runs immediately to the executive level. And the answer, in the absence of a governance record, is the same answer it has always been: whoever had the organizational authority to know and did not build the structure to find out.

The reason this problem has gone unsolved in most organizations is simpler than it appears: there has been no shared language for describing what Shadow AI is, where it comes from, or how to find it. When the technology team says "we don't have a Shadow AI problem," they typically mean: we have not received any IT tickets for unauthorized AI deployments. That is not the same thing. The absence of a ticket does not mean the absence of a system. It means the system arrived without generating one. The five categories below give you the taxonomy to find what the ticket queue will never show you.

THE SHADOW AI TAXONOMY

Shadow AI does not arrive in one form. It arrives in five. Each category has a different origin, a different risk profile, and a different detection method. Understanding all five is what separates an AI inventory that captures your actual exposure from one that captures only what someone thought to declare.

For any organization with more than fifty employees, all five categories are almost certainly present. The question is not whether. The question is what, specifically, and where.

> ***TYPE 1, INDIVIDUAL SUBSCRIPTIONS***
>
> *These are AI tools purchased directly by employees, often on personal or corporate cards, occasionally expensed, almost never reviewed by IT or legal. They include writing assistants, research tools, image generators, presentation builders, code assistants, and the rapidly expanding category of AI agents that can take actions on behalf of a user. The defining characteristic of this category is that the purchasing decision and the deployment decision are made by the same person, at the same moment, with no organizational review in between.*

HOW IT ARRIVES: *A team member finds a tool that solves a problem. The monthly cost is low enough to charge on a corporate card without approval. They start using it. They invite colleagues. Within ninety days, a workflow dependency has formed. The tool is now embedded in how a team operates, and no one with governance authority has ever seen it.*

DETECTION:

A spend analysis of corporate card transactions filtered for SaaS and subscription vendors, cross-referenced against the approved software list. Any vendor not on the approved list whose product description includes AI capability is a candidate for review. This analysis can typically be completed by a finance analyst in one to two days.

TYPE 2, EMBEDDED AI IN EXISTING SAAS TOOLS

This category is the most insidious, because it does not require any employee action to activate. It arrives when a software vendor your organization already uses quietly deploys AI capabilities inside their product. Email platforms, CRM systems, HR software, document management tools, financial systems, and project management applications have all introduced AI features, often enabled by default, often processing data that flows through those systems automatically, without explicit notification to the customer's IT department, legal team, or governance function. Your organization did not purchase AI. It purchased a contract. The AI arrived in an update.

HOW IT ARRIVES: *Vendor product updates, often released quarterly or more frequently, enable new AI features within existing subscriptions. Unless the organization has a formal process for reviewing software updates against the data processing terms of their vendor agreements, these capabilities activate and begin processing organizational data without review.*

DETECTION:

A systematic review of product update logs and release notes for your ten highest-data-volume SaaS vendors over the past twenty-four months. Focus on updates describing AI, machine learning, automation, or 'intelligent' features. For each: when did this activate in our instance, what data does it process, and did our data processing agreement with this vendor cover this use? This review should involve both IT and legal.

TYPE 3, DEVELOPER-BUILT INTERNAL MODELS

In any organization with a technical staff, software engineers, data analysts, data scientists, or technically proficient business analysts, there is a meaningful probability that someone has built an AI model for internal use without formal governance review. These range from simple machine learning scripts that score leads or forecast inventory to more complex models that make recommendations or automate decisions. The common thread is that they were built by individuals or small teams to solve specific operational problems, without going through the formal system approval process that governs enterprise software deployments.

HOW IT ARRIVES: *A technically skilled employee identifies a problem that a predictive model can solve. They build it, test it on internal data, and deploy it for their team's use. It works. Other teams ask to use it. It scales informally. By the time anyone in governance has visibility, it has become a de facto decision-support tool the business depends on, one that has never been reviewed for bias, data governance compliance, or accuracy under distribution shift.*

DETECTION:

A survey distributed to all technical staff, engineering, data, analytics, and operations functions, asking: have you built or maintained any automated or AI-assisted tool used by people other than yourself for business decisions? Keep the survey short, anonymous if necessary for candor, and framed explicitly as an inventory exercise, not a compliance investigation. Chapter Four covers the full AI asset register; this survey is the first-pass detection mechanism for this category.

TYPE 4, VENDOR-PROVIDED AI IN EXISTING CONTRACTS

This category covers AI capabilities contractually included in vendor agreements your organization already holds, but either not understood to be AI at the time of signing, or significantly expanded since the original contract was executed. Managed service providers, consulting firms, staffing vendors, logistics partners, and financial services providers have increasingly embedded AI into their service delivery without making this explicit in contract language. Your vendor is using an AI model to do the work your contract specifies. You may not know. Your contract may not address it. Your governance structure almost certainly does not cover it.

HOW IT ARRIVES: *Vendors incorporate AI into their operations to improve efficiency, accuracy, or speed. The output you receive, a report, a recommendation, a scored data file, a processed claim, may now be the product of an AI model rather than a human analyst, without disclosure or contract amendment. The liability transfer questions this creates are significant: whose model was it, what data did it process, what were its outputs used for, and who is accountable if those outputs were wrong?*

DETECTION:

A targeted review of your twenty largest vendor contracts, asking two questions of each: (1) Does the contract specify whether AI will be used in service delivery, and if so, what governance rights does your organization retain? (2) In the past twelve months, has this vendor's service delivery changed in ways that suggest AI augmentation, faster turnaround, different output formats, new 'insights' features? This review should be conducted by your legal or procurement function with explicit AI governance framing.

TYPE 5, API INTEGRATIONS BY TECHNICAL STAFF

The fifth category requires the least capital and leaves the fewest traces. A technically skilled employee, an engineer, a developer, a data analyst with programming ability, can connect an external AI service to an internal system through an Application Programming Interface in an afternoon. The connection routes organizational data to an external model for processing and returns results. From the inside, it looks like a feature. From the outside, it is a data transfer to a third-party AI system your organization has not evaluated, whose data processing terms your legal team has not reviewed, and whose model behavior is entirely outside your governance perimeter.

HOW IT ARRIVES: *An engineer or analyst wants to add an AI capability to an internal tool or workflow. They sign up for API access, write a few lines of code to route data through the external service, and deploy the integration. It may never appear in a software procurement review because no software was purchased, only API calls were made, billed at a per-use rate that accumulates on a cloud services bill.*

DETECTION:

A review of cloud services billing, AWS, Azure, GCP, or equivalent, for API call charges associated with AI services. Additionally, a review of code repositories for integration code referencing external AI provider endpoints. This detection method requires coordination between finance and IT, and is typically more productive when framed as a cost governance exercise than a compliance investigation.

Five categories. Five different arrival paths. Five different detection methods. No single audit surfaces all of them. If you have not run these five detection methods, you do not have a complete picture of the AI operating in your organization. You have a picture of the AI someone thought to tell you about.

Those are not the same picture.

EIGHTEEN MONTHS WITHOUT A TICKET

Illustrative Scenario: *The following scenario is representative of documented patterns in executive AI governance. Names, organizations, and specific details are illustrative.*

The operations VP at a mid-size manufacturing company, approximately $2.4 billion in annual revenue, publicly traded, discovered their Shadow AI problem during a routine vendor review. Not an AI governance review. A vendor consolidation project whose purpose was reducing the number of software subscriptions the company was paying for.

The project was being led by a procurement analyst who was three months into the role. She was building a master list of every software subscription the company held. When she got to the ERP system, the company's core operational platform, through which virtually all supply chain, inventory, and production planning data flowed, she pulled the current subscription terms and compared them against the original contract.

There was a feature she did not recognize. A module called Demand Intelligence, described in the current documentation as an "AI-powered inventory forecasting and procurement optimization engine."

She searched the original contract. No mention. She searched the amendment log. No amendment. She asked the IT manager responsible for the ERP system when this module had been activated.

He checked. It had been activated automatically, as part of a standard product update, eighteen months prior.

For eighteen months, the Demand Intelligence module had been ingesting the company's full inventory data, supplier relationship data, production schedules, and historical procurement pricing and using it to generate purchasing recommendations that the procurement team had been following as standard practice. Nobody had reviewed the module's data processing terms. Nobody had evaluated whether the vendor's use of that data, including, as the current terms specified, the right to use aggregated customer data for model improvement, was consistent with the company's supplier agreements and competitive sensitivity requirements. Nobody had assessed whether the recommendations the module was generating were being followed with sufficient human review to satisfy the organization's own procurement policy.

The operations VP, when she was briefed, asked three questions in sequence. The first was: how long has this been running? Eighteen months. The second was: does our legal team know? No. The third was: could a competitor, through the vendor's aggregated data, gain insight into our pricing strategy?

Nobody had an immediate answer to the third question.

The legal review that followed took six weeks. The findings were mixed: the vendor's data use terms were aggressive but not clearly in breach of the company's supplier agreements. The competitive sensitivity risk was real but difficult to quantify. The procurement team's reliance on the module's recommendations without documented human review was inconsistent with the company's own procurement policy. The operations VP deactivated the module pending a formal governance review, negotiated amended data processing terms with the vendor, implemented a documented review protocol for AI-generated procurement recommendations, and added the module to the company's AI asset register, which, prior to this exercise, did not exist.

The exposure had been running for eighteen months before anyone with governance authority knew it existed. It was not discovered by an AI governance audit. It was discovered by a procurement analyst doing a different job, who compared a current contract against an original one and noticed a feature she did not recognize. Detection was accidental.

The detection method the taxonomy specifies for Type 2, a review of product update logs for your highest-data-volume vendors, would have found this system within days of activation. It did not exist. So the module ran quietly for eighteen months, processing data that touched competitive positioning, supplier relationships, and procurement policy. The organization found out by accident.

"Nothing has gone wrong" is a status report. It is not a governance record. A system that processes sensitive organizational data and informs significant business decisions requires oversight regardless of how it arrived, regardless of whether the vendor meant well, and regardless of whether a problem has surfaced yet.

BUILDING THE MAP BEFORE MONDAY

The five categories above are the foundation of your Shadow AI detection effort. Four actions you can take immediately that will give you a materially more accurate picture of your Shadow AI exposure than you currently have, before the full AI inventory process in Chapter Four begins.

1. **Run the spend analysis.**

Ask your finance team, today, not at the next planning cycle, to pull a list of every SaaS and subscription vendor charged on corporate cards or through accounts payable in the past twelve months. Cross-reference that list against your approved software vendor list. Every vendor not on the approved list is a candidate. Filter for any whose product name or description includes AI, machine learning, automation, intelligence, or assistant. That filtered list is your first-pass Type 1 Shadow AI inventory. In most organizations with five hundred or more employees, this exercise surfaces between eight and twenty-five unapproved AI tools within forty-eight hours.

Send this to Finance, not IT. Your IT department sees what generated a ticket. Finance sees everything that generated a charge. The spend analysis is faster, broader, and does not depend on employee self-disclosure. Success looks like a list, imperfect and incomplete but real, of AI tools operating in your organization that no one in a governance role has ever reviewed.

2. **Audit your five largest SaaS vendors for AI activation.**

Identify the five software vendors through whom the greatest volume of organizational data flows. For each one, pull the current product documentation or release notes from the past twenty-four months and scan for AI-related feature releases. Then ask IT to confirm: is this feature active in our instance? When was it activated? What data does it process? You do not need to review all your vendors at once. The five highest-data-volume vendors represent the majority of your Type 2 exposure, and this targeted audit can typically be completed in a week.

Do not ask the vendor. Check their release notes independently. Vendors confirm what you ask about. They do not proactively disclose what you fail to ask about. The update log review is verification, not a trust exercise. Success looks like a clear answer, for each of your five largest vendors, to the question: is AI running in this system, processing our data, and if so, under what terms?

3. **Send the internal developer survey.**

Draft a short, direct survey to all technical staff, engineering, data, analytics, and operations functions. Three questions: Have you built or maintained any automated or AI-assisted tool used by people other than yourself for business purposes? If yes, describe it briefly and identify who uses it. Do you know of any AI tools or integrations built by colleagues that are not on this list? Make the survey anonymous if your culture requires it for candor. Frame it explicitly as an inventory exercise, not a compliance investigation, because if your technical staff believe they are being audited for wrongdoing, you will get incomplete answers. The goal is a list, not a judgment.

Send it directly to individual contributors, not through technology leadership. Managers filter based on what they believe should have been built. Individual contributors report what was. Make it anonymous if your culture requires candor, and put the non-judgmental framing in the first sentence. Success looks like a list of internal AI tools that no one in governance had documented, built by people who were solving problems, not creating exposure.

4. **Brief your General Counsel before you have all the answers.**

Do not wait until the inventory is complete to involve your legal team. Brief your General Counsel now, with whatever partial picture you have, on two things: first, that you are conducting a Shadow AI detection exercise and will be producing a formal inventory; and second, that based on what you know, you believe there may be data processing activities and vendor AI deployments that have not been reviewed for legal compliance. This brief accomplishes two things simultaneously. It creates a record that you initiated governance review, which matters under the 'should you have known' standard from Chapter One. And it activates your legal team as a governance partner rather than a last resort.

Most common mistake: treating legal involvement as a sign that something has gone wrong. In the AI governance context, early legal involvement is a sign that something is going right. The executive who brings their General Counsel in at the beginning of a Shadow AI review is building a governance record. The executive who brings them in after a regulator asks the first question is managing a crisis. Success looks like a shared understanding, between you and your legal team, of what the exercise is, why it matters, and what documentation will be produced.

THE MAP IS THE FIRST GOVERNANCE ACT

The operations VP in this chapter did not discover her organization's Shadow AI through vigilance. She discovered it through accident. The eighteen months it took to surface were eighteen months of unreviewed data processing, unexamined competitive exposure, and ungoverned decision support, none of which appeared in any governance record because the governance structure did not exist to generate one. The system was not malicious. It was invisible. Invisible AI carries exactly the same legal exposure as visible AI, without any of the governance documentation that might demonstrate reasonable oversight.

The single most important insight in this chapter is this: you cannot govern what you have not mapped, and the map does not build itself. It requires a deliberate, structured detection effort across all five Shadow AI categories, because each category arrives differently and each requires a different method to surface.

> ***Shadow AI is not a technology problem your IT department failed to solve. It is a governance vacuum your leadership structure has not yet filled. The four actions above are where you start filling it.***

The four actions above are a first-pass detection effort, not a complete inventory. They will give you a meaningfully more accurate picture of your Shadow AI exposure by the end of next week. The complete inventory process, the one that produces the formal AI asset register your board can review and your governance program can be built on, is Chapter Four.

But before you can govern the AI you have, you need to understand precisely why governing it is your obligation and not your technology team's. The fiduciary reframe that makes everything in this book actionable is the subject of Chapter Three. It is also, in my experience, the chapter that changes how executives think about their role in the age of AI more permanently than any other.

The map is not the governance program. It is the foundation without which the governance program cannot stand.

BOARDROOM TAKEAWAY | *The Fiduciary Reframe*

What a board member should be able to ask:

- What governance structure exists to ensure our AI systems are doing what we have authorized them to do?
- For each material AI system, who is the named fiduciary accountability owner, and is that documented in their role description?

What a CFO or risk leader should have documented:

- A written identification of three material AI systems with named accountability owners for each
- A legal counsel memo, however preliminary, on your organization's current AI governance fiduciary exposure

What evidence should exist:

- A revised board or committee agenda that includes AI Governance Status as a standing item, separate from AI capability updates
- A written record that fiduciary accountability for AI governance has been claimed explicitly, not assumed

CHAPTER THREE: PART I: THE RECKONING

The Fiduciary Reframe

Why This Is Your Problem, Not IT's

This chapter makes one argument. It is not new. It is four centuries old, embedded in the foundations of Anglo-American corporate law, and it applies to you right now in a domain you almost certainly believe belongs to your technology team.

The argument is this: the fiduciary obligation to govern AI is not a new standard being created by emerging regulation. It is an existing standard, Duty of Care, Duty of Loyalty, Business Judgment Rule, that has been applied to executive behavior across every material domain of organizational risk for decades. AI is now a material domain. The obligation transferred the moment it became one. The story that follows is about what happens when an executive discovers this too late, and what happens when one discovers it just in time.

> **Illustrative Scenario:** *The following scenario is representative of documented patterns in executive AI governance. Names, organizations, and specific details are illustrative.*

The CEO still has his job. The board asked him to stay.

The CTO, who had built the AI program, who had championed every deployment, who had presented the capability roadmap at three consecutive board meetings with slides that described a competitive moat and a productivity multiplier, the CTO is gone. She left by mutual agreement fourteen months after an AI system her team deployed began producing recommendations that a subsequent audit found were systematically disadvantaging a protected class of customers. The enforcement action named the organization. The board investigation named her. The CEO's name appeared in the findings only once, in a passage that noted he had not been provided adequate governance information by his technology leadership.

That sentence, had not been provided adequate governance information, does two things at once. In one reading, it exonerates the CEO. He was not informed. He relied on the people whose job it was to inform him. In another reading, the reading that will increasingly matter as courts and regulators develop their AI jurisprudence, it is a description of a governance failure for which the CEO bears as much responsibility as the CTO. The obligation to be informed is not passive. It does not discharge when the right briefing fails to arrive. For a fiduciary, the obligation to be informed about material organizational risk is affirmative. It does not wait for the briefing. It demands the briefing.

The CEO's survival in that story was not the product of innocence. It was the product of timing, board relationships, and the availability of someone closer to the decision to hold responsible. Those are not governance tools. They are luck. And in the AI accountability environment that is now developing, luck has a shorter half-life every year.

This chapter makes the case you have been hoping would not need to be made: this is your responsibility, it has always been your responsibility, and the question is not whether you will be held to a fiduciary standard for AI governance. The question is whether you will build the structure that lets you meet it.

THE DELEGATION REFLEX AND WHY IT FAILS

Every executive reading this book has, at some point in the last three years, done some version of the same thing. An AI-related question came up, in a board meeting, a staff meeting, a regulatory examination, a vendor negotiation, and you turned to the person with the technical title. The CTO. The CIO. The Chief Data Officer. The VP of Engineering. You turned to them not because you were abdicating responsibility but because the question felt technical, and technical questions have a natural gravitational pull toward the people whose job title suggests they understand the technology.

This is the delegation reflex. It is not unique to AI. It is how organizations function in every domain where specialized expertise is required. You do not personally review every legal contract, you have a General Counsel. You do not personally audit every financial statement, you have a CFO, or you are the CFO and you have an audit team. You build structures, hire capable people, and govern through accountability rather than personal mastery of every domain.

The delegation reflex works in every domain where the governance structure has been built to support it. It fails in AI because that structure does not exist.

> ***The delegation reflex works in every domain where the governance structure has been built to support it. In AI, for most organizations, that structure does not yet exist.***

The cost of this gap compounds in a way that most executives have not fully priced. Each quarter that passes without a governance structure is a quarter in which AI systems operate without documented oversight, in which board reporting on AI risk is capability-framed rather than governance-framed, and in which the 'should you have known' standard from Chapter One compounds. The question is not whether the standard applies. It does. The question is whether, when it is eventually applied to your organization, there will be a governance record that demonstrates you met it, or an absence of record that demonstrates you did not build the structure required to meet it.

The reason capable executives have not solved this problem yet is not inattention or indifference. It is that the fiduciary case for AI governance has not been made to them in terms they recognize as their own. The AI governance conversation has been conducted in the technology department's language, models, data pipelines, accuracy rates, bias metrics, and it has not landed in the executive suite as a fiduciary obligation because nobody has translated it into the language that fiduciaries work in. This chapter is that translation.

THE THREE FIDUCIARY OBLIGATIONS APPLIED TO AI

Fiduciary duty is not a new concept. It is one of the oldest obligations in corporate law, and senior executives have been operating under its requirements, in one form or another, for as long as they have held positions of organizational authority. The framework is stable. The application to AI is new. And the translation from one to the other is simpler than the AI governance industry has made it appear.

There are three fiduciary obligations that directly govern how executives must approach AI: the Duty of Care, the Duty of Loyalty, and the Business Judgment Rule. Each has been applied to executive behavior in corporate law for decades. Each now applies, as a matter of existing legal doctrine and emerging regulatory expectation, to AI governance. Understanding what each obligation requires, and what each means for the specific practices of governing AI, is the reframe that makes everything in the chapters ahead not optional but necessary.

OBLIGATION 1, THE DUTY OF CARE

You must inform yourself about material risks. AI is now material.

The Duty of Care requires that a corporate fiduciary, in making decisions that affect the organization, exercise the level of care, diligence, and skill that a reasonably prudent person in a similar position would exercise. In practice, this means that before making a decision of consequence, you are obligated to inform yourself, to gather the information a reasonable person would gather, to ask the questions a reasonable person would ask, and to seek the expertise a reasonable person would seek. The duty is not satisfied by receiving a briefing. It is satisfied by receiving an adequate briefing and acting on it. The Delaware courts, whose corporate law jurisprudence sets the benchmark applied across US jurisdictions, have held that a director or officer who makes decisions in a state of willful ignorance, who deliberately avoids information that a reasonable person would have sought, cannot claim the protection that attaches to informed business judgment.

AI GOVERNANCE IMPLICATION:

AI systems that make decisions affecting customers, employees, pricing, credit, hiring, or resource allocation are, by any reasonable definition, material to the organization's risk profile. The Duty of Care therefore requires that you inform yourself: what AI systems are operating, what they are deciding, what risks they create, and what governance structure surrounds them. The marketing manager's individual subscription from Chapter Two is not material. The AI underwriting model that processes forty thousand loan applications is. The line between material and immaterial is not always obvious, which is exactly why Chapter Five builds the risk tiering framework. But the obligation to draw that line, and to inform yourself adequately about what sits above it, is yours.

OBLIGATION 2, THE DUTY OF LOYALTY

You must act in the organization's interest, including managing the risks it does not yet see.

The Duty of Loyalty requires that a fiduciary act in the best interests of the organization and its stakeholders rather than in their own personal interest or in the interest of a preferred constituency. In the AI context, this obligation is less often about personal enrichment, the classic conflict-of-interest scenario, and more often about the subtle conflict between the organization's short-term operational convenience and its long-term governance health. When an AI system produces valuable operational results but creates unreviewed legal exposure, the Duty of Loyalty requires that the executive address the exposure rather than accept the results and defer the review. When a vendor relationship is commercially important but the vendor's AI deployment has not been evaluated under the organization's governance standards, the Duty of Loyalty requires the evaluation rather than the commercial convenience of not conducting it.

AI GOVERNANCE IMPLICATION:

The Duty of Loyalty, applied to AI, is the obligation to govern AI in the organization's genuine long-term interest even when doing so creates short-term friction. The AI program that is delivering impressive ROI metrics and has never been reviewed for fair lending compliance is not an asset being protected by avoiding the review. It is a liability being grown. The executive who defers the governance work to preserve the operational momentum is, in the framework of fiduciary law, placing convenience above obligation. This is the doctrine that makes deferral a fiduciary issue rather than merely a strategic one.

OBLIGATION 3, THE BUSINESS JUDGMENT RULE

You are protected when you make informed decisions. You lose that protection when you delegate without oversight.

The Business Judgment Rule is the fiduciary doctrine that most executives know best, even if they do not know it by name. It holds that courts will not second-guess a business decision made by a fiduciary who was adequately informed, acted in good faith, and had no personal conflict of interest. It is the legal recognition that executives must make decisions under uncertainty, that not every decision will be correct, and that the legal system should not substitute its judgment for the executive's in the ordinary conduct of business. The protection is real and substantial. When it applies, executives are shielded from personal liability for decisions that turn out badly. When it does not apply, when the decision was made in a state of inadequate information, or when the executive had delegated without maintaining meaningful oversight, that protection disappears and personal liability becomes a live question.

AI GOVERNANCE IMPLICATION:

The Business Judgment Rule is the single most important reason to build the governance structure in this book. Not because it eliminates AI risk, but because it determines whether you are personally protected when that risk materializes. The executive who governed AI with documented oversight, adequate information, and a board-level accountability structure can point to the governance record and claim the Rule's protection. The executive who delegated to the technology team without maintaining meaningful oversight cannot. The Rule does not protect delegation. It protects informed, good-faith decision-making. Those are not the same thing, and the difference between them is the governance record this book builds.

Three obligations. One governing principle: fiduciaries are accountable for what they should have known, not only for what they were told. That principle has always been true.

AI makes it urgent.

> ***The Standard in Plain Language***
> *You are not required to become a technologist. You are required to govern like a fiduciary. That means being adequately informed about material AI risks, acting in the organization's genuine long-term interest in managing them, and making documented, good-faith decisions that the Business Judgment Rule can protect. The tools to do all three are built in the chapters that follow.*

The discipline the Rule requires, decisions made on adequate information, with risks named and alternatives weighed, is not unique to AI. It is the same discipline that protects every consequential executive choice, and applying it beyond AI is the work of *The Decision Balance Sheet.*

THE TECHNOLOGY COMMITTEE THAT CHANGED ITS MIND

> **Illustrative Scenario:** *The following scenario is representative of documented patterns in executive AI governance. Names, organizations, and specific details are illustrative.*

A publicly traded company in the professional services sector had been managing AI governance the way most organizations manage it: as a technology function responsibility with periodic board reporting on AI initiatives, framed primarily around investment and capability.

The board's technology committee received a quarterly briefing from the CTO. The briefing was well-prepared, clearly presented, and consistently optimistic. It covered the AI tools the organization had deployed, the efficiency gains they had produced, the new capabilities under development, and the vendor relationships that supported the program. It did not cover governance. Not because the CTO was hiding anything. Because governance had not been requested. The committee had asked for a technology briefing, and a technology briefing was what they received.

In the third quarter of what I will call Year One of this story, a new member joined the technology committee. She had spent nine years as a general counsel at a financial institution and had lived through two regulatory examinations in which AI-related governance deficiencies had been cited. She did not know more about machine learning than the other committee members. She knew more about what a regulator looks for when it examines an organization's AI governance posture.

At her first technology committee meeting, after the CTO's quarterly briefing, she asked one question: "This briefing describes what our AI program can do. Can you tell the committee what governance structure exists to ensure that what it can do is what we have decided it should do, and that we would be able to demonstrate that to an examiner?"

The CTO paused. It was not an evasive pause. It was a genuine one, the pause of a technically sophisticated person encountering a question that fell outside the frame of reference they had been operating in. She said she could prepare a response for the next meeting.

The committee chair, who had been on the board for six years and had approved the technology committee's AI briefing format without modification, asked a follow-up: "Should we have been asking this question before now?"

The new committee member said: "Yes. But you're asking it now, and that matters."

What followed was eight months of governance construction that the technology committee drove and the technology team executed. The committee adopted a formal AI governance charter that established board-level oversight as a standing function rather than a periodic briefing. It commissioned an AI asset register, the first the organization had ever produced, which identified twenty-seven AI systems, seven of which had never been formally approved by any governance function. It established a risk tiering protocol that classified each system by decision autonomy and consequence magnitude. It adopted a board reporting template, built on the format in Chapter Eight of this book, that replaced the capability briefing with a governance summary the entire board could read and interrogate.

When the organization's primary regulator conducted a routine examination eighteen months later, the examination team spent two days reviewing the AI governance program. Their written findings noted that the organization's AI oversight structure was "consistent with emerging supervisory expectations" and that the board's active engagement with AI governance demonstrated "an appropriate exercise of fiduciary oversight." No deficiencies were cited. No corrective actions were required.

The committee member did not know more about AI than her colleagues. She knew which question to ask, and she understood it was her obligation to ask it. The Duty of Care required her to seek adequate information. The Duty of Loyalty required her to act on what it revealed. The Business Judgment Rule protected the organization when the regulator arrived because the documentation of good-faith governance existed to meet the standard.

The question could have been asked two years earlier, by anyone with governance authority who understood their obligation. It was not. Not because those people were bad fiduciaries. Because nobody had given them the reframe.

You are not the committee that failed to ask this question for three years. But you were in the same position as every person in that room who had governance authority and did not yet know which question to ask. That changed with this chapter. You now have it.

CLAIMING THE OBLIGATION BEFORE IT CLAIMS YOU

The fiduciary reframe is not academic. It changes what you do on Monday. Five actions that translate the three obligations into immediate practice, each one a governance move rather than a technology move, executable without a single line of code or a single consultation with your technology team.

Before Part II begins, this chapter closes with one question, not a list. Not five actions, not a delegation to your legal team, not a governance checklist. One question, to be answered in writing, personally, today.

Take the new committee member's question and adapt it for your specific context: what governance structure exists to ensure that our AI systems are doing what we have decided they should do, and that we could demonstrate responsible oversight to an examiner? Write that question down, in your own words, as if you were going to ask it at your next board or risk committee meeting. The act of writing it will tell you immediately whether you know the answer. If you do not know the answer, that is your Duty of Care gap. The question itself, asked in the right meeting, begins to close it.

Most common mistake: softening the question to avoid the discomfort of not having an answer. The question is not comfortable. That is the point. A governance question that everyone can answer with confidence is a question about a governance structure that already exists. The question above will not be answerable with confidence in most organizations, and the discomfort that produces is the organizational signal that the governance work needs to begin. Success looks like asking the question aloud, in a governance context, and creating a record that you asked it.

What is the first AI governance obligation you have been treating as someone else's responsibility that is, under the fiduciary standard this chapter describes, yours?

The Duty of Care requires that you be adequately informed about material AI risk. Materiality is not infinite. You are not required to govern every AI system with equal intensity, the risk tiering framework in Chapter Five will calibrate that. But you are required to know, personally, what the highest-stakes AI decisions in your organization are: the systems making credit decisions, pricing decisions, hiring decisions, clinical recommendations, resource allocations, or any other consequential determination that affects people's access to opportunity or resources. Name three of them. Write them down. For each one, write a single sentence describing who, in your current organizational structure, is the named accountability owner for that system's governance. If you cannot write that sentence, the accountability chain for that system does not exist.

Materiality in the fiduciary sense tracks consequence, not cost. An AI system that cost two million dollars to build but generates marketing recommendations carries a different risk profile than an AI system that cost forty thousand dollars but determines which loan applications advance to underwriting. Name the most consequential systems, not the most expensive ones. Success looks like three named systems, three named accountability owners, and the honest recognition of whether those owners have been given governance authority or merely technical responsibility.

Write the answer down. That is where the structure in Part II begins.

For each of the three AI systems you identified in step two, ask yourself the question a court would ask if that system's outputs were challenged: if I were required to testify that I made informed, good-faith decisions about the governance of this system, what documentation would I produce? A board briefing that addressed governance rather than capability? A risk assessment that established the basis for the deployment decision? An oversight protocol that shows ongoing monitoring? A vendor contract that specifies accountability terms? For each system, for each document type, you either have it or you do not. The gap between what you have and what you would need is your Business Judgment Rule exposure. The chapters that follow close those gaps one by one.

A three-line email approving a vendor invoice is a document. It is not governance documentation. The Business Judgment Rule requires evidence of an informed decision-making process: adequate information gathered, risk assessed, alternatives considered, conclusion reached and defensible. The specific documents that constitute that record are built in Chapters Four through Eight. For now, name the gap between what you have and what you would need to produce under examination. That gap is your governance workplan.

If you have a board-level or executive-level technology briefing scheduled in the next sixty days, revise the agenda before it occurs. Add one agenda item: AI Governance Status. Define that agenda item explicitly as distinct from the AI capability update: it covers what systems are operating, what governance documentation exists for each, what the current risk posture is, and what oversight mechanisms are in place. Send the revised agenda to whoever prepares the briefing with a brief note: we are adding a governance section, separate from the capability update, that addresses these four questions. The note itself is a governance act. It creates a record that you requested a governance briefing, which demonstrates the exercise of Duty of Care before the briefing has even occurred.

Keep them separate and label them separately. A governance section folded into the technology briefing produces exactly the problem the new committee member identified: a briefing that describes what the AI program can do, with no account of what governance structure exists to ensure it does only what the organization has authorized. The distinction is not bureaucratic. It is fiduciary. Success looks like a board or committee that receives both types of information, understands the difference, and begins to expect both as a matter of standard quarterly reporting.

Schedule thirty minutes with your organization's top legal officer. General Counsel, Chief Legal Officer, or equivalent, and ask them directly: in their assessment, what is the current state of the organization's legal exposure related to AI governance? Not AI regulation broadly. Not the technology department's compliance posture. Executive fiduciary exposure, under the three obligations in this chapter, given what they know about how AI is currently governed in the organization. This conversation creates a record that you sought legal counsel on this specific question before taking governance action, a Duty of Care demonstration. And it activates your legal team as a governance partner rather than a last resort.

Your General Counsel may not have a fully formed view when you ask. Most legal teams, unless specifically directed to assess AI governance fiduciary exposure, have been tracking AI regulation and contract issues rather than this specific question. Your question will prompt a research exercise. That is correct. What matters is that you asked it, on a specific date, and that the conversation is in the record. Success looks like a written summary, however preliminary, of your General Counsel's assessment of your current AI governance exposure: a document that exists and is dated before the governance program begins.

THE QUESTION IS NOT WHETHER. IT IS WHETHER YOU ARE READY.

The CEO in the opening of this chapter survived his organization's AI enforcement action. He survived it because of timing, board relationships, and the availability of someone closer to the decision to hold responsible. That is not a governance strategy. It is not a defense. And it is not a model for how you want to be positioned when the standard is eventually applied to your organization's AI governance.

The new committee member in this chapter protected her organization not by knowing more about AI than her colleagues but by knowing which question to ask and understanding it was hers to ask. That question flows directly from the three obligations described in this chapter. The Duty of Care required her to ask it. The Duty of Loyalty required her to act on the answer. The Business Judgment Rule protected the organization when the regulator arrived, because the documentation of good-faith governance existed.

The single most important insight in this chapter is this: the fiduciary obligation to govern AI is not a new standard being created by emerging regulation. It is an existing standard. Duty of Care, Duty of Loyalty, Business Judgment Rule, being applied to a new domain. You already knew how to meet this standard in every other material risk area of your organization. The reframe is not learning something new. It is recognizing something that has always been yours to own.

> ***The duty to govern AI is not coming. It is here. The question is not whether you are responsible. The question is whether you are prepared.***

Part I of this book is complete. Three chapters. Three realities established: the legal exposure exists and is documented; the AI you do not know about carries the same liability as the AI you do; and the governance obligation is yours under doctrine that has been settled law for decades. You cannot unknow these things. You should not want to. Because the executive who has absorbed Part I and acted on it is the executive who arrives in Part II with the right frame, not a technology manager trying to understand AI, but a fiduciary building a governance structure that the law already expects them to have.

Part II begins with the most foundational governance act in this entire book: building the AI inventory. Not commissioning a technology audit. Not asking your IT department to compile a software list. Building the formal AI asset register that gives your governance program its foundation, gives your board the visibility the Duty of Care requires, and gives the Business Judgment Rule the documentation it needs to protect you.

You have the obligation. You have the reframe. Part II builds the structure.

CHAPTER FOUR: PART II: THE FRAMEWORK

Knowing What You Have

Before you read this chapter, answer one question.

How many AI systems is your organization currently running? Not a rough estimate. The actual number, from a document that exists.

If a document does not exist, that is the answer this chapter is about. Because what follows is not a story about a CFO who did not care about AI governance. It is a story about what it costs, precisely, to not have that document when the board asks the question.

The AI Inventory Your Board Expects

She had been CFO for four years when the board chair asked the question she could not answer.

It was not a hostile question. It was not even an unusual one, given the direction board conversations about AI had been trending for the prior eighteen months. The chair leaned forward in her seat, looked down the table, and said: "Before we approve the next tranche of AI investment, I'd like to understand what we're already running. Can someone give me a list of every AI system currently operating in this organization, who owns each one, and what it's authorized to decide?"

> **Illustrative Scenario:** *The following scenario is representative of documented patterns in executive AI governance. Names, organizations, and specific details are illustrative.*

The CFO looked at the CTO. The CTO looked at his notes. The Chief Risk Officer opened her laptop. The silence lasted eleven seconds, the CFO counted, before someone said they would need to pull that together and have it ready for the next meeting.

The board chair nodded. Made a note. Moved on.

In the car afterward, the CFO called the CTO. She said: "That question is going to keep coming up. We need a document that answers it before it's asked again." The CTO agreed. He assigned it to a senior analyst on his team. The analyst spent three weeks pulling data from IT ticketing systems, software procurement records, and department budget reports. She produced a spreadsheet with thirty-one rows.

The CFO later told me it was the most clarifying document she had seen in four years in the role. Not because the list was short, thirty-one systems was more than she had expected. Not because every system on the list was well-governed, several were not. But because for the first time in her tenure, she knew exactly what her organization was running, what each system was authorized to do, and who was responsible when it did something wrong. She knew because the document existed. And the document existed because a board chair had asked an eleven-second question that nobody could answer.

This chapter is about building that document before you need to find out in a board meeting that you do not have it.

THE DOCUMENT THAT DOES NOT EXIST

In three years of conversations with senior executives about AI governance, I have asked a version of the same question to hundreds of people in positions of organizational authority: do you have a single document that lists every AI system operating in your organization, what each one is authorized to decide, and who is accountable for it? The consistent answer, across industries, organizational sizes, and levels of general governance sophistication, is some variation of no.

What follows is almost always the same: most executives believe that the information exists somewhere. They believe their IT department has a software inventory. They believe their technology team tracks the AI programs under active development. They believe their procurement function has records of vendor contracts that include AI components. And they are generally right. All of those things exist, in fragments, in separate systems, owned by separate functions, formatted in ways that were never designed to produce a governance document. The information is distributed. The document is not assembled.

> ***The information exists in fragments. The document has not been assembled. And a governance record that has not been assembled is, for all practical purposes, a governance record that does not exist.***

The cost of this gap shows up in three places. First, in board conversations like the one that opened this chapter: the organization cannot demonstrate to its own board that it knows what it is running. Second, in regulatory examinations: examiners in financial services, healthcare, and increasingly in other regulated industries ask for AI asset documentation as a standard examination request, and organizations that cannot produce it receive findings that reflect the governance gap regardless of whether any system has actually malfunctioned. Third, and most expensively, in litigation: when an AI system is challenged, the first discovery request typically asks for documentation of what the system was, what it was authorized to do, and who governed it. The organization that cannot produce that documentation does not merely lose the documentation question. It loses the governance narrative.

The reason this document does not exist in most organizations is organizational rather than negligent. No one function owns the complete picture. IT owns the technical inventory. Procurement owns the contract records. Business lines own the deployment decisions. Legal owns the compliance assessments, when they exist. Finance owns the cost records. The AI asset register is the document that assembles those fragments into a single governance record, and it does not exist because no one has been given the governance mandate to build it. That mandate comes from you. This chapter gives you the template.

THE AI ASSET REGISTER: FIVE COLUMNS, ONE GOVERNANCE RECORD

The AI asset register is a single document, a structured list, that captures every AI system operating in your organization with enough detail to support governance decisions at the board, executive, and operational levels simultaneously. It is not a technology audit report. It is not a software inventory. It is not a project tracking spreadsheet. It is a fiduciary record: evidence that your organization knows what AI it is running, what authority each system holds, and who is accountable for each one.

The register has five columns. Each answers a specific governance question. Together they produce a record that satisfies the Duty of Care standard from Chapter Three, provides the foundation for the risk tiering process in Chapter Five, and gives your board the visibility the Business Judgment Rule requires.

COL 1 SYSTEM / VENDOR	COL 2 BUSINESS FUNCTION	COL 3 DATA INPUTS & OUTPUTS	COL 4 DECISION AUTHORITY	COL 5 ACCOUNTABLE OWNER	COL 6 TIER
LoanSense AI Acme Lending Corp	Consumer loan application scoring	IN: Application data, credit bureau, income verification OUT: Score, approval recommendation	Autonomous, outputs route applications without human review below threshold	Chief Lending Officer	3
Salesforce Einstein Salesforce Inc.	Sales opportunity scoring and next-action recommendations	IN: CRM activity, deal history, contact data OUT: Opportunity score, recommended actions	Advisory, sales reps receive recommendations; all decisions require human approval	VP of Sales	1
HireView Screen HireView Inc.	Initial candidate screening and interview scheduling	IN: Résumés, application responses, job requirements OUT: Ranked candidate list, screen/no-screen recommendation	Assistive, output informs recruiter decision; human confirmation required to advance or reject	Chief People Officer	2
Demand Intelligence [ERP Vendor, embedded, not separately contracted]	Procurement optimization and inventory forecasting	IN: Inventory levels, supplier data, production schedules, pricing history OUT: Purchase recommendations, reorder triggers	Autonomous, recommendations execute automatically below defined thresholds	[UNASSIGNED, requires governance action]	3*

*Tier key: 1 = Advisory (low autonomy, low stakes) ■ 2 = Assistive (moderate autonomy, moderate stakes) ■ 3 = Autonomous (high autonomy, high stakes) ■ * = requires immediate governance review*

Five columns. Every AI system in your organization gets a row. The template above shows four illustrative examples, including one shadow AI system, shown in amber, that carries no assigned owner and requires immediate governance action. That row represents what the spend analysis and vendor audit from Chapter Two surface: a system that is running, that no one has formally claimed, and that cannot be governed until it has an owner. The register does not just inventory what is governed. It makes visible what is not.

Exhibit 1: AI Asset Register

EXHIBIT 1: AI ASSET REGISTER					
Column 1 System Name / Vendor	Column 2 Business Function	Column 3 Data Inputs / Outputs	Column 4 Decision Authority	Column 5 Accountable Owner	Tier (1/2/3)
CreditScore AI / VendorCo	Consumer loan underwriting, supplements underwriter review	In: applicant financials, credit history, employment Out: risk score + recommendation	Assistive: AI drives the recommendation; underwriter approves or overrides	Chief Credit Officer [Name]	2
HR Screening Tool / WorkflowAI	Applicant screening for open positions, filters to shortlist	In: résumés, job descriptions, historical hiring data Out: ranked shortlist; candidates below threshold auto-rejected	Autonomous at screening stage: candidates rejected without human review	SVP Human Resources [Name]	3
Meeting Summarizer / NoteAI	Generates summaries of internal meetings; distributed to attendees	In: meeting audio/transcript Out: summary document; no decision authority	Advisory only: human reviews and distributes summary	IT Operations [Name]	1
Fraud Detection / SecureAI	Real-time transaction monitoring, flags and holds suspicious activity	In: transaction data, account history, behavioral patterns Out: block/hold decision executed automatically	Autonomous: system blocks transaction without human pre-approval	Chief Risk Officer [Name]	3

Instructions: One row per AI system. Complete all five governance columns before assigning tier. Tier is determined by decision autonomy and consequence level, not cost or visibility. Update quarterly. Retain as a governance record.

COLUMN 1, SYSTEM NAME & VENDOR

WHY IT MATTERS: *You cannot govern a system you cannot name. Column 1 is the identity record: the specific name of the AI system or tool, the vendor who provides it (or the internal team that built it, for developer-built models), and whether it arrived through formal procurement or through one of the shadow AI pathways in Chapter Two.*

WHAT TO RECORD: *The system's commercial name or internal designation, the providing vendor or builder, the contract or procurement reference if one exists, and the date the system was first deployed in the organization.*

COMMON PITFALL:

Recording only vendor names without system names, or recording only the platform (e.g., 'Salesforce') without identifying the specific AI module within it. The governance record needs to be specific enough that a new executive who has never worked at the organization can look at Column 1 and know exactly what system is being discussed.

COLUMN 2, BUSINESS FUNCTION

WHY IT MATTERS: *Every AI system exists to serve a business purpose. Column 2 records that purpose in plain language, not the technical description of how the system works, but the organizational description of what it does and for whom. This column is most likely to surface a mismatch between what the system was deployed to do and what it is actually being used for.*
WHAT TO RECORD: *The department or function that uses the system, the specific business process it supports, and the primary output the business relies on. One to two sentences in plain language. If the system's actual current use has drifted from its original deployment purpose, note both.*
COMMON PITFALL:
Describing the system's technical function rather than its business function. 'Natural language processing model for text classification' is a technology description. 'Sorts inbound customer complaints by category and routes to the appropriate service team' is a governance description. The register should contain the latter.

COLUMN 3, DATA INPUTS & OUTPUTS

WHY IT MATTERS: *This column most directly enables the data governance and privacy compliance assessment. AI liability, particularly in the discriminatory outputs and data privacy categories from Chapter One, attaches to what data a system processes and what it produces. Column 3 creates the record that makes a compliance review possible without starting from scratch.*
WHAT TO RECORD: *The categories of data the system ingests (not individual data fields, but data categories: customer personal data, financial transaction data, employee performance data, health data, behavioral data, etc.) and the form of the system's outputs (a score, a recommendation, a classification, an automated action). Flag any inputs that fall into protected or sensitive categories: health information, demographic data, financial data, or behavioral data that could serve as a proxy for protected characteristics.*
COMMON PITFALL:
Leaving this column blank because the answer requires technical consultation. If you do not know what data a system processes, that absence of knowledge is itself a governance finding, and it belongs in the register as a flagged gap rather than an empty cell. A gap you can see is a gap you can close.

COLUMN 4, DECISION AUTHORITY

WHY IT MATTERS: *This is the single most important column for governance purposes, because it determines the system's risk tier and the oversight intensity it requires. Decision authority describes whether the system advises, assists, or acts autonomously. The distinction is not technical, it is organizational. The same model*

technology can function as advisory in one deployment and autonomous in another, depending on how the business has configured its workflows around it.
WHAT TO RECORD: *One of three designations: Advisory (the system produces a recommendation or analysis that a human uses as one input among many, with no expectation of compliance), Assistive (the system produces a recommendation that a human is expected to follow unless they have a specific reason to override it), or Autonomous (the system produces an output that triggers an action or routes a decision without requiring human confirmation). If the system's actual operation does not match its designed authority level, if it was deployed as advisory but functions as autonomous in practice because the override rate is below five percent, record the actual operational authority, not the designed one.*
COMMON PITFALL:
Recording the designed authority level without verifying actual operational authority. A system classified as advisory in its deployment documentation may function as autonomous in practice if the business line follows its outputs without review. The governance register must reflect operational reality. If you do not know the override rate for a consequential system, that is a governance gap that belongs in the register as a flagged unknown.

COLUMN 5, ACCOUNTABLE OWNER

WHY IT MATTERS: *Every AI system must have a named human being, by role, not just by function, who is the organizational accountability owner for that system's governance. Not the person who built it. Not the vendor who provides it. The internal executive or senior leader whose role explicitly includes being accountable for the system's outputs, its compliance posture, and its governance documentation.*
WHAT TO RECORD: *The specific role title of the accountable owner (Chief Lending Officer, Chief People Officer, VP of Operations, etc.). Not a department name and not a committee. One person whose job description, in writing, includes accountability for this system. If no such person currently exists for a given system, record the gap explicitly: '[UNASSIGNED, requires governance action]' in amber, as the shadow AI example in the register above shows. An unassigned system is a governance emergency, not an empty row.*
COMMON PITFALL:
Naming the CTO or CIO as the accountable owner for every system. The technology leader may be the right technical escalation point, but they are not the appropriate fiduciary accountability owner for systems whose outputs affect customers, employees, or external parties. The accountable owner should be the executive who carries organizational responsibility for the domain the system operates in: the Chief Lending Officer for a credit scoring model, the Chief People Officer for a hiring tool, the CFO for a financial forecasting system.

> ***The Tier Column***
> *The sixth column. Tier, is assigned automatically based on Column 4. Advisory systems are Tier 1 (lower oversight intensity). Assistive systems are Tier 2 (moderate oversight intensity). Autonomous systems are Tier 3 (highest oversight intensity, requiring explicit board-level visibility). The full risk tiering methodology is in Chapter Five. For now, assign the tier based solely on the decision authority designation. The tier drives everything downstream: how often the system is reviewed, what governance documentation it requires, how it appears in the board report, and what insurance and liability reserve considerations it triggers.*

THIRTY-ONE SYSTEMS IN TWENTY-THREE DAYS

The regional bank completed its first AI asset register in twenty-three days. I want to walk through how they did it, because the process they used is instructive for organizations that are starting from the same place most organizations start: distributed information, no assembled document, and a governance function that has not previously owned this work.

The CFO, the same one from the chapter's opening, whose board chair had asked the question that could not be answered, did not assign the register project to IT. She formed a four-person cross-functional task force: one person from IT, one from legal, one from procurement, and one from finance. She gave the task force a single deliverable: a completed five-column register for every AI system currently operating in the organization, using the template above. She gave them twenty-three business days. She made herself the executive sponsor, which meant she received a status update every five days and was the decision authority on any classification dispute.

The task force used three detection passes. The first was the spend analysis from Chapter Two: corporate card transactions and accounts payable records filtered for SaaS and subscription vendors, cross-referenced against the approved software list. This pass surfaced fourteen systems. The second was a targeted review of the five largest data-volume vendors' product update logs for the prior two years, looking for AI feature activations. This pass surfaced six more, including an embedded AI module in the bank's core banking platform that had been activated eleven months earlier without anyone in governance having reviewed it. The third was a short survey to all technical staff asking whether they had built or maintained any automated tools used by others for business decisions. This pass surfaced eleven more, including two internally built models that the compliance team had never reviewed.

Thirty-one systems. Seven had never been through any form of formal governance review. Three had no named accountable owner. Two had Column 4 designations that did not match their actual operational behavior: both were classified as advisory in their approved documentation, but follow-up interviews with the business lines confirmed that the recommendations were followed without human review in more than ninety percent of cases, making them functionally autonomous. The task force reclassified both to Tier 3 and flagged them for the CFO.

On Day Twenty-Three, the CFO presented the completed register to the board's risk committee. The committee chair, the same board chair who had asked the original question, spent forty minutes reviewing it. She asked seven specific questions, all of which the register was designed to answer. Three of those questions were about the Tier 3 systems: what was the oversight protocol, who was the named accountable owner, and what would trigger an escalation to the full board. The CFO had answers to all three, because the register had required her to build the answers in order to complete it.

The board chair said: "This is the most useful governance document I've seen since we implemented the risk committee.

I want to see it quarterly."

The register did not just tell the board what was running. It forced the organization to resolve governance gaps that had existed, unnoticed, for months or years, unassigned owners named, misclassified systems reclassified, undisclosed embedded AI disclosed and governed. A register is not a reporting document. It is a governance catalyst: a structure that makes invisible problems visible and forces accountable ownership onto every system in the portfolio.

That is what a fiduciary record does. It does not merely document what is known. It surfaces what cannot be ignored.

BUILDING YOUR REGISTER IN THIRTY DAYS

The bank's twenty-three-day timeline is achievable because the task force was focused and the executive sponsor was engaged. Four actions that launch the process and keep it moving.

1. **Form the task force today.**

The register cannot be built by one function acting alone, because no one function has the complete picture. You need four seats at the table: IT (for the technical inventory), legal (for contract and compliance review), procurement (for the vendor spend analysis), and finance (for cost records and budget cross-referencing). Appoint yourself as executive sponsor, not as a participant. Your role is to set the mandate, receive status updates, and make classification decisions when the task force disagrees. The task force does the research. You provide the governance authority that makes the work non-negotiable.

Four to six people is optimal. Every additional person adds coordination overhead without proportionally adding detection capability. The goal is speed, not comprehensiveness at this stage. Success looks like a task force with named members, a confirmed Executive Sponsor, and a first meeting scheduled before the end of the week.

2. **Run all three detection passes in the first ten days.**

The task force should complete the spend analysis, the vendor update review, and the technical staff survey in the first ten days of the thirty-day window. Running all three passes before beginning to populate the register ensures that the register captures the full population of systems rather than the obvious ones. The most governance-significant systems are often the ones the technology team did not think to volunteer. The spend analysis and vendor review surface them without requiring anyone to admit they exist.

Most common mistake: populating the register with known systems before completing the detection passes, then treating the detection passes as optional. The instinct is to make visible progress quickly by documenting the AI everyone already knows about. Resist it. A register that captures twenty of thirty-one systems is not a governance record, it is a partial inventory with a false sense of completion. Run the detection passes first. Success looks like three completed detection lists merged into a single master list before Column 1 is populated for a single row.

3. **Conduct a one-hour classification workshop for every Tier 3 candidate.**

For every system whose Column 4 designation is Autonomous, or whose actual override rate suggests it is functionally autonomous regardless of its designed designation, conduct a one-hour classification workshop with the relevant business line and the task force. The workshop answers four questions: What is the system actually deciding? What data is it processing? Who, by name, is the accountable owner? And what governance documentation currently exists for this system? The workshop does not need to produce a governance plan. It needs to produce completed register rows with confirmed classifications and named owners. One hour per system, scheduled in the first fifteen days.

Column 4 is a governance question, not a technical one. The technology team can describe how a system operates. Determining whether that operation constitutes Advisory, Assistive, or Autonomous decision-making is an executive judgment with legal implications. Own the classification. Success looks like a Column 4 entry for every system that reflects actual operational behavior, not the vendor's marketing description or the technology team's self-assessment.

4. **Present a draft register to your board's risk committee before it is complete.**

Do not wait for a perfect, fully populated register before bringing it to the board. Present a draft at thirty days, even if some rows have flagged gaps, unassigned owners, or pending classification reviews. A draft register with visible gaps is more valuable as a governance demonstration than a polished register produced six months later. It shows the board that the process is active, that the organization is taking the inventory seriously, and that the executive team is building the structure rather than waiting for an external prompt. The gaps you present are not weaknesses, they are the governance work in progress, and their presence in a board document demonstrates that someone is looking.

A governance register is never complete. It is current. Present it to the board when it is current, not when it is perfect. An incomplete register presented at Day 30 with a completion plan is stronger governance evidence than a completed register presented at Day 90 with no board visibility during construction. Success looks like a board committee that receives the register before it is final, understands what it represents, and is positioned to ask governance questions about what it contains.

THE FOUNDATION EVERYTHING ELSE STANDS ON

The CFO from this chapter's opening told me, eighteen months after that eleven-second silence in the board meeting, that the AI asset register was the single most consequential governance document her organization had produced in the prior decade. Not because it was the most sophisticated. Because it was the most foundational. Everything that followed, the risk tiering, the accountability chain, the board reporting, the vendor negotiations, the incident response protocol, was built on the register. Without it, those structures had no substrate. With it, they had a map to build from.

The single most important insight in this chapter is this: the AI asset register is not a technology document. It is a fiduciary record, evidence that your organization knows what it is running, what authority it has granted to each system, and who is accountable when that authority produces a consequence. Every governance structure in the chapters ahead assumes this record exists. Without it, you are governing by assumption. With it, you are governing by knowledge. The difference between those two postures is the difference between the executive who testifies and the one who produces the document.

Your register will show you that you are running multiple AI systems with different levels of decision authority, different data sensitivities, and different consequence profiles. Governing all of them with the same intensity is neither efficient nor strategically sound. You need a way to allocate your governance attention, to concentrate oversight where the risk is highest and calibrate it appropriately where it is lower. That is the work of Chapter Five.

The register is the first governance act. It is not the last. An inventory that is never tiered by risk produces the illusion of governance without the substance of it: a list of systems, each receiving the same level of oversight regardless of whether it recommends a playlist or makes an autonomous credit decision. Chapter Five solves that problem. It is the chapter most organizations skip, to their significant cost.

BOARDROOM TAKEAWAY | *Risk That Isn't Abstract*

What a board member should be able to ask:

- Are any of our Tier 3 autonomous systems operating without a named executive accountability owner?
- Have we verified whether any system classified as advisory is functioning as autonomous in actual practice?

What a CFO or risk leader should have documented:

- A completed risk tier classification for every system in the AI Asset Register
- Override rate data for every Tier 1 and Tier 2 system, verifying that operational behavior matches designed authority

What evidence should exist:

- A written governance record showing the two-dimension tier assessment for each system
- Escalation documentation for any system reclassified from Tier 1 or 2 to Tier 3 based on operational reality

CHAPTER FIVE: PART II: THE FRAMEWORK

Risk That Isn't Abstract

How to Grade, Tier, and Own Your AI Exposure

Two AI systems. One policy. One catastrophic misallocation of governance attention.

The organization had done something most organizations have not: they had written an AI governance policy. It ran to fourteen pages. It specified review cycles, documentation requirements, vendor assessment protocols, and escalation thresholds. It had been drafted by legal, approved by the board, and distributed to all technology leaders. By any conventional measure, it represented serious governance effort.

The policy applied uniformly to every AI system in the organization. The same documentation requirements. The same review cycle. The same escalation threshold. The same governance intensity for a content recommendation engine that suggested which blog posts to feature on the company website as for the underwriting model that determined whether small business owners received the credit they needed to keep their companies alive.

> **Illustrative Scenario:** *The following scenario is representative of documented patterns in executive AI governance. Names, organizations, and specific details are illustrative.*

Eighteen months after the policy was adopted, a regulatory examination found that the underwriting model had been producing decisions with a statistically significant demographic pattern for fourteen of those months. The governance review cycle required by the policy was annual. The model had been reviewed once, at implementation. The annual review was scheduled for month fifteen. The examination occurred in month fourteen.

The organization had a governance policy. What they did not have was a governance policy calibrated to risk. They had treated a content recommendation engine and an underwriting model as equivalent governance problems because their policy was built around the category 'AI system' rather than the question that actually matters: what is this system deciding, for whom, and what happens to real people when it decides wrong?

The answer to that question is not the same for every system. This chapter gives you the framework to calibrate your governance to the answer.

THE UNIFORM TREATMENT TRAP

The instinct to treat all AI systems with the same governance protocol is understandable. It feels equitable. It feels systematic. One policy, one review cadence, one documentation standard, applied consistently across the portfolio. That instinct is wrong, and acting on it produces one of two governance failures.

The first failure is under-governance of high-risk systems. When the review cycle for an autonomous credit-scoring model is set at the same frequency as a blog recommendation engine, the model's actual risk profile, the magnitude of its decisions, the populations it affects, the legal exposure it creates, is not receiving the oversight it requires. The governance intensity is calibrated to the average system in the portfolio rather than the most consequential one. In a portfolio where the most consequential system carries liability exposure orders of magnitude above the median, that miscalibration is not a minor inefficiency. It is a fiduciary failure waiting for a regulatory examination to surface it.

The second failure is over-governance of low-risk systems. When every AI system in the portfolio receives the full governance treatment, formal risk assessments, legal reviews, quarterly board reporting, vendor audit protocols, regardless of what it is actually deciding, the governance function burns resources on systems that do not require them. The result is governance fatigue: the organization's capacity for rigorous oversight is consumed by low-stakes systems, leaving the high-stakes systems nominally covered but practically under-scrutinized.

> ***Uniform governance produces the appearance of oversight without the substance of it. Tiered governance concentrates the substance where the stakes are highest.***

The reason most organizations have not solved this problem is that tiering AI risk requires a classification framework that most governance functions do not have. The legal team knows how to assess contract risk. The technology team knows how to assess model performance risk. Neither is trained to assess the compound risk that matters for fiduciary governance: the combination of how autonomously a system acts and how consequentially it acts. That compound assessment requires neither technical expertise nor legal training. It requires two honest questions about each system in the register, and the discipline to allocate oversight based on the answers.

THE AI RISK TIER MODEL: TWO DIMENSIONS, THREE TIERS

The AI Risk Tier Model classifies every system in your AI asset register into one of three governance tiers based on two dimensions. The first dimension is decision autonomy: how much of the decision does the AI make on its own, without requiring human confirmation before an action occurs? The second dimension is consequence magnitude: how significantly does the output of this system affect the people, resources, or obligations it touches? Every AI system in your organization sits somewhere on each of these two dimensions. The intersection of where it sits determines its tier.

The model is deliberately simple. It gives you a consistent, defensible, executive-level classification that determines how much governance attention each system receives, how often it is reviewed, how it appears in board reporting, and what insurance and liability reserve considerations it triggers. It is the framework that turns the AI asset register from an inventory into a governance priority map.

Exhibit 2: AI Risk Tier Model

EXHIBIT 2: AI RISK TIER MODEL

		DECISION AUTONOMY		
		Advisory (Human decides; AI informs)	Assistive (AI drives; human approves)	Autonomous (AI decides; no prior approval)
CONSEQUENCE LEVEL	Low (Internal / reversible outcomes)	TIER 1 Advisory Standard monitoring	TIER 1 Advisory Standard monitoring	TIER 2 Assistive Enhanced oversight
	High (External / irreversible / regulated outcomes)	TIER 1 Advisory Standard monitoring	TIER 2 Assistive Enhanced oversight	TIER 3 Autonomous Executive oversight required

TIER	Governance Requirements	Review Cadence	Board Visibility	Override Rate Threshold
1	Named accountable owner. Documented purpose and data inputs. Annual review.	Annual	Portfolio summary only	Monitor; no threshold required
2	Named executive sponsor. Documented accountability chain (Links 1-5). Semi-annual review. Override rate tracked.	Semi-Annual	Named in risk section	Flag if below 15% in consequential domain
3	Named C-suite executive sponsor. Full 6-link chain documented and confirmed. Quarterly board reporting. Tabletop exercise required. Incident protocol in place.	Quarterly	Named individually in every board summary	Escalate if below 5%, reclassify as fully autonomous

Classification rule: When decision autonomy and consequence level point to different tiers, classify at the higher tier. Reclassify based on actual override rates, not designed intent.

	LOW DECISION AUTONOMY	HIGH DECISION AUTONOMY
HIGH CONSEQUENCE	**TIER 2** **Assistive** *Human decides, AI drives*	**TIER 3** **Autonomous** *AI decides, human monitors*
LOW CONSEQUENCE	**TIER 1** **Advisory**	**TIER 2** **Assistive**

	Human decides, AI informs	*Moderate autonomy, assess context*

Classification rule: *When the two dimensions point to different tiers, classify at the higher tier. Governance intensity should match the worst plausible outcome, not the average expected one.*

The asymmetry in the matrix is deliberate. A system with high decision autonomy and low consequence magnitude is still Tier 2, not Tier 1, because autonomous systems can make many decisions quickly, compounding low-magnitude errors into high-magnitude patterns before anyone with oversight authority has noticed. The content recommendation engine is not dangerous because any single recommendation matters enormously. It becomes dangerous when it systematically recommends content that disadvantages a particular group, at scale, without anyone monitoring the aggregate pattern.

Here is what each tier requires across the four governance dimensions:

TIER 1, ADVISORY AI
The system informs. A human decides.

Tier 1 systems produce outputs, recommendations, analyses, predictions, rankings, that a human uses as one input among many. The human making the decision is not expected to follow the AI's recommendation. They may follow it, disregard it, or weigh it against other inputs. The defining characteristic is genuine human discretion: the AI advises, and the human decides independently. Examples include marketing analytics tools, research assistants, content suggestion engines, scheduling optimization tools, and diagnostic support systems where the human professional retains full decision authority.

GOVERNANCE REQUIREMENTS
A named accountable owner in the AI asset register. Annual documentation review confirming the system's advisory status has not drifted toward assistive or autonomous use in practice. Vendor contract reviewed for data processing terms at implementation and at each renewal. Basic incident log maintained by the accountable owner.

REVIEW FREQUENCY
Annual review by the accountable owner. No scheduled board-level review unless a material change in use case, data processing, or vendor terms occurs.

BOARD REPORTING OBLIGATION
Included in the AI governance summary as an aggregate count. No individual Tier 1 system requires board-level reporting unless it crosses into Tier 2 territory.

INSURANCE CONSIDERATION
Standard technology errors-and-omissions coverage is typically sufficient for Tier 1 systems. Verify with your broker that your current policy covers AI-generated advisory outputs. No dedicated liability reserve required for individual Tier 1 systems.

TIER 2, ASSISTIVE AI
The human decides, but the AI drives the decision.

Tier 2 systems produce outputs that a human is expected to follow unless they have a specific, articulable reason to override. The distinction from Tier 1 is not technical, it is organizational and behavioral. When the default assumption is compliance with the AI's recommendation rather than independent human judgment, the system is operating as Tier 2 regardless of how it was designed or how the policy describes it. The practical test: what is the override rate? If fewer than twenty percent of the AI's recommendations are overridden in normal operation, the system is functioning as Tier 2 at minimum. Examples include clinical decision-support tools with low override rates, hiring screening systems whose rankings determine interview pools, credit pre-qualification tools, fraud alert systems that automatically block transactions, and pricing tools whose outputs set published prices.

GOVERNANCE REQUIREMENTS
Named accountable owner with documented governance authority. Semi-annual review of system outputs for accuracy, bias indicators, and demographic pattern analysis. Documented override protocol specifying when and how humans can and should reject the AI's recommendation. Vendor contract reviewed for retraining notification, performance accountability, and data use terms. Bias and fairness assessment completed at implementation and annually thereafter.

REVIEW FREQUENCY
Semi-annual review by the accountable owner, with results reported to the relevant executive. Annual summary included in the board AI governance report.

BOARD REPORTING OBLIGATION
Individual Tier 2 systems in high-consequence domains (credit, employment, healthcare, pricing) are reported individually in the board AI governance summary with their override rate, most recent bias assessment result, and accountable owner. Tier 2 systems in lower-consequence domains are reported as an aggregate.

INSURANCE CONSIDERATION
Review your current technology and professional liability coverage specifically for AI-assisted decision liability. Most standard policies require an AI endorsement for Tier 2 coverage. Discuss dedicated liability reserve with your CFO and insurance broker, sized to the consequence magnitude of the domain the system operates in.

TIER 3, AUTONOMOUS AI
The AI decides. Human oversight is structural, not transactional.

Tier 3 systems make decisions and take actions without requiring human confirmation on each individual output. A human may set the parameters, monitor the aggregate behavior, and have the authority to intervene or shut the system down, but no human is reviewing and approving each decision before it occurs. The consequence of this architecture is that errors scale at machine speed before oversight can catch them. The fourteen-month underwriting pattern in this chapter's opening was not discovered by the organization's governance protocol. It was discovered by a regulatory examination. Tier 3 systems require governance intensive enough to catch systematic errors before the regulator does. Examples include autonomous credit scoring models, algorithmic trading systems, automated content moderation at scale, clinical decision systems with automatic order execution, dynamic pricing engines that set prices without human approval, and any system that takes consequential action on behalf of the organization without per-transaction human review.

GOVERNANCE REQUIREMENTS

Named executive-level accountability owner whose role explicitly includes Tier 3 governance. Quarterly output review with statistical sampling for accuracy, bias, and demographic pattern analysis. Documented human oversight protocol specifying the monitoring cadence, escalation triggers, and intervention authority. Vendor contract with explicit retraining notification, performance SLA, incident disclosure obligation, and data ownership terms. Pre-deployment fairness and impact assessment. Documented decision log retained for the period required by applicable regulatory standards. Board-level visibility at every governance reporting cycle.

REVIEW FREQUENCY

Quarterly review by the accountable owner, with written summary submitted to the relevant board committee. Any material finding, bias indicator, accuracy degradation, demographic pattern, vendor incident, triggers immediate board notification outside the standard cycle.

BOARD REPORTING OBLIGATION

Every Tier 3 system is reported individually in the board AI governance summary at every reporting cycle. The report includes: system name and function, most recent output review result, override and exception rate, any incidents since last report, and accountable owner's assessment of current risk posture. The board should be able to name every Tier 3 system in the organization from memory after two reporting cycles.

INSURANCE CONSIDERATION

Tier 3 systems require dedicated coverage review. Engage your insurance broker specifically on AI autonomous decision liability, which is treated differently than technology errors-and-omissions in most current policies. Establish a dedicated liability reserve for each Tier 3 system, sized in consultation with your CFO and legal team based on the consequence magnitude of the domain and the volume of decisions the system makes. This reserve is part of the total cost of AI ownership analysis in Chapter Six.

The Operational Reality Test

Classify every system based on its actual operational behavior, not its designed behavior or policy designation. The question is not: what tier was this system deployed as? The question is: what tier is it operating as today? A system deployed

> *as Tier 1 whose recommendations are followed without review ninety percent of the time is a Tier 2 system with a Tier 1 label.*
> ***The label does not determine the exposure. The behavior does.***

THE CLINICAL TOOL THAT RECLASSIFIED ITSELF

The health system had been using a clinical decision-support tool in its emergency department for two years. The tool analyzed patient intake data, vitals, chief complaint, triage assessment, medical history, and produced a clinical pathway recommendation: which tests to order, which specialists to consult, and in some cases, whether to admit or discharge. The tool had been reviewed at implementation by the clinical informatics team and classified as advisory. The physician retained full decision authority. The tool informed. The physician decided.

That was the designed architecture. The operational reality had quietly become something different.

> **Illustrative Scenario:** *The following scenario is representative of documented patterns in executive AI governance. Names, organizations, and specific details are illustrative.*

In the first six months of deployment, the override rate for the tool's recommendations was approximately thirty-one percent. Physicians frequently modified the suggested pathway, added tests the tool had not recommended, or disregarded the discharge recommendation in favor of admission. The tool was being used as one input among many, as designed.

By the end of the second year, the override rate was four percent.

Nobody had changed the policy. Nobody had changed the tool's classification in the asset register. Nobody had issued guidance telling physicians to follow the tool's recommendations more closely. What had happened was subtler: the tool had a strong accuracy record. Physicians who overrode its recommendations and were subsequently shown by outcomes to have been wrong experienced the normal human recalibration toward the more reliable input. The tool's recommendations became the presumed correct path. Override became the exception that required justification.

The tool had drifted from advisory to autonomous without a single policy change, a single vendor update, or a single governance decision.

The governance gap was surfaced by a quality review that examined patient outcomes by pathway recommendation. The review found that the tool's recommendation for a specific chief-complaint category, atypical cardiac presentations, had a systematic underperformance pattern for a specific demographic group. Because the override rate was four percent, that underperformance had propagated through approximately ninety-six percent of cases in that category for an unknown duration. The clinical informatics team could not determine precisely when the drift in override rate had begun, because no one had been monitoring it.

The tier model would have caught this. A Tier 1 classification with a quarterly operational reality check would have flagged the override-rate drift by month eighteen at the latest. That triggers reclassification to Tier 3, which triggers quarterly output reviews with statistical sampling. The underperformance surfaces in a review rather than a quality investigation, and months earlier.

The health system's governance failure was not deploying an AI tool without a policy. It was deploying a tool with a static classification in a dynamic operational environment. AI systems do not stay in their initial tier. Physician behavior changes. Override rates shift. Vendor updates alter model behavior. Organizational workflows evolve to accommodate the tool's outputs. The tier model is not a one-time classification exercise. It is an ongoing operational assessment, revisited at every review cycle, because the question 'what tier is this system operating as today?' does not have a permanent answer.

TIERING YOUR REGISTER THIS WEEK

The tier classification does not require a technology assessment, a regulatory review, or an external consultant. It requires honest answers to two questions for each system in your AI asset register. Four actions that complete the tier assignment and activate the governance protocols each tier requires.

1. **Apply the two-dimension test to every system in your register.**

For each system in your AI asset register, answer two questions in writing. First: on a scale from fully advisory to fully autonomous, where does this system actually operate, based on the observed override rate, not the policy designation? If you do not know the override rate for a consequential system, that is your first action item: find out before assigning the tier. Second: if this system made the same error one thousand times in a row before anyone noticed, what would the consequence be for the people or resources it touches? Low consequence means the error is annoying or inefficient. High consequence means the error affects someone's access to credit, employment, healthcare, housing, financial resources, or legal standing. The intersection of your two answers determines the tier.

Answer the consequence question based on the worst plausible output, not the average. The tier classification protects against the bad day, not the typical day. If the worst plausible output of a system affecting credit, employment, healthcare, or housing is consequential for the person affected, classify at the higher tier regardless of how infrequently that output occurs. Success looks like a tier classification for every system that your General Counsel would support if it appeared in a regulatory examination.

2. **Run the operational reality test on every Tier 1 and Tier 2 system.**

Pull the actual override or exception rate for every system currently classified as Tier 1 or Tier 2 in your register. For Tier 1 systems operating in consequential domains, any override rate below twenty percent should trigger reclassification review. For Tier 2 systems, any override rate below five percent in a high-consequence domain should trigger reclassification to Tier 3. The override rate is not the only signal, changes in how the tool is integrated into workflows, whether outputs are reviewed before or after action, and whether the business line treats the recommendation as presumptively correct are all relevant. The point is not a bright-line rule. It is an honest operational assessment that the tier reflects how the system is actually being used.

Most common mistake: accepting the accountable owner's characterization of the override rate without verifying it against system logs or outcome data. Accountable owners are often unaware of how thoroughly their teams have come to rely on the AI's outputs. The clinical informatics team in this chapter's story did not know the override rate had fallen to four percent until the quality review surfaced it. Verify against data, not against self-report. Success looks like a confirmed override rate for every system in the register, with any system whose rate has fallen below the reclassification threshold formally reviewed and either reclassified or documented as having been assessed and confirmed at its current tier.

3. **Assign the governance protocol for each tier immediately.**

Once the tier assignment is complete, the governance protocol for each system follows from the tier card above, it is not a separate design exercise. For Tier 1 systems: confirm the named accountable owner and set the annual review date in their calendar today. For Tier 2 systems: confirm the named accountable owner, schedule the semi-annual output review, and verify that the vendor contract contains the retraining notification and performance accountability terms. For Tier 3 systems: confirm the executive-level accountable owner, schedule the quarterly output review with written reporting, and add each Tier 3 system as a standing item on the board's risk committee agenda. The protocol is not a future project. It activates with the tier assignment.

The governance requirements on the tier card are not aspirational. They are the baseline. A Tier 3 system without a full six-link accountability chain is not a Tier 3 system with incomplete governance. It is an ungoverned system with a Tier 3 risk profile. The tier card tells you what is required. Success looks like every Tier 3 system fully compliant with its tier requirements before the first Board AI Governance Summary is presented.

4. **Add tier to the board AI governance summary.**

In the next board AI governance report, whether that is a document you build from scratch in Chapter Eight or a modification of what you are already producing, include a tier distribution summary: the number of systems at each tier, the names of all Tier 3 systems, and the accountable owner for each. This single addition transforms a capability briefing into a risk briefing. The board can see not just that AI is running but how much of it is operating autonomously in high-consequence domains, who is responsible for each one, and what the governance cadence is. That is the information a board needs to exercise the oversight the Duty of Care requires.

Name every Tier 3 system individually in every board presentation. A board that knows the organization has "two Tier 3 systems" is receiving portfolio information. A board that knows the organization is running CreditScore AI and SecureAI, their accountable owners, and their current override rates, is receiving governance information. The specificity is the governance. Success looks like a board that can name the organization's highest-risk AI systems from memory after three consecutive quarterly presentations.

CALIBRATED GOVERNANCE IS ACCOUNTABLE GOVERNANCE

A governance policy that applies the same intensity to every AI system in the portfolio is not rigorous governance. It is the appearance of governance, systematic, equitable, and wrong. The fourteen-month underwriting pattern and the four-percent override rate both existed inside organizations that had written documentation, named owners, and review schedules. What they did not have was a governance framework that matched the oversight intensity to the actual risk. That is what the tier model provides.

The single most important insight in this chapter is this: tier your AI systems based on what they are doing, not what you intended them to do, and revisit the classification at every review cycle because the operational reality shifts without anyone deciding to change it.

Calibrated governance is not a concession to resource constraints. It is the correct governance posture: intensive oversight for systems that warrant it, proportionate oversight for systems that do not, and the documented rationale that distinguishes one from the other.

The TCAIO model in Chapter Six adds the financial dimension to the picture the tier model builds: not just how risky each system is, but what each one actually costs.

> ***Tier your risk and you can govern it. Treat it uniformly and you will eventually be governed by it.***

BOARDROOM TAKEAWAY | *The Real Price Tag*

What a board member should be able to ask:

- What is the actual total cost of our AI portfolio across all seven TCAIO categories, not just the subscription line items?
- Has our board approved AI investment with full visibility into compliance, liability, and governance overhead costs?

What a CFO or risk leader should have documented:

- A TCAIO cost model covering all seven categories for at least the highest-cost AI systems
- A comparison of the total AI cost estimate against the currently approved AI budget

What evidence should exist:

- A documented TCAIO estimate for every system above a defined deployment threshold
- Board or committee documentation showing that AI investment decisions were made with full-cost visibility

CHAPTER SIX: PART II: THE FRAMEWORK

The Real Price Tag

Every CFO knows the difference between the cost of a thing and the total cost of a thing. It is one of the first distinctions a financial education teaches: the subscription price is not the operating cost, the acquisition price is not the carrying cost, the licensing fee is not the governance cost. You have applied this discipline to capital equipment, to real estate, to human resources, to debt instruments.

Most organizations have not applied it to AI. Here is what that costs.

Total Cost of AI Ownership Beyond the Subscription

The CFO approved a $340,000 annual AI investment. She was proud of the number. It was disciplined. It was documented. It had survived a rigorous business case review that showed a 2.8-times return on investment within eighteen months. The board approved it unanimously. The technology team celebrated.

Fourteen months later, I sat across from her in a conference room and asked her to walk me through the actual cost of that AI program. Not the budget line. The actual cost: what the organization was spending, in aggregate, on every expense category the AI program had created or expanded, including the ones that did not appear in the technology budget.

> **Illustrative Scenario:** *The following scenario is representative of documented patterns in executive AI governance. Names, organizations, and specific details are illustrative.*

The exercise took two hours. When we were done, the number on the whiteboard was not $340,000. It was $1.2 million.

She stared at it for a long time. Then she said something I have heard in some version in nearly every organization I have worked with on AI cost modeling: "I knew the subscription wasn't the whole story. I didn't know the rest of the story was three and a half times the subscription."

The 2.8-times return on investment her board had approved was calculated against $340,000. Against the actual cost of $1.2 million, the program was generating a 0.8-times return. It was cash-flow negative. The organization was losing money on an AI investment its board believed was among its best capital allocation decisions of the year.

Nobody had lied. Nobody had manipulated the business case. The $340,000 was real. The costs that brought the total to $1.2 million were also real, they simply existed in organizational categories that the AI business case framework was not designed to capture: data infrastructure, compliance staffing, legal review cycles, insurance adjustments, executive governance time, and the liability reserve the CFO's own risk committee had quietly established for the organization's two Tier 3 systems without anyone connecting it to the AI ROI calculation. The business case was not wrong. It was incomplete.

An incomplete cost model is not a financial analysis. It is a wish list with a spreadsheet attached.

This chapter builds the complete model.

WHAT THE P&L CANNOT SEE

Every organization that has deployed AI has a visible AI cost: the licenses, the subscriptions, the vendor contracts, the implementation fees. These numbers appear somewhere in the technology budget, and they are the numbers that governance functions, board presentations, and ROI analyses are built around. They are also, in almost every organization I have examined, a minority of the actual cost of running AI responsibly.

The majority of AI's true cost lives in organizational categories that were not designed to capture it. Data infrastructure costs grew because the AI system needed more storage, faster processing, and cleaner data pipelines, but those costs were absorbed into the IT budget as a general infrastructure expansion, not attributed to the AI program. Compliance costs increased because legal needed to review the vendor contract amendments, assess the fair lending implications of the new underwriting model, and prepare responses to regulatory inquiries, but those costs were tracked in the legal budget as standard professional services. Executive time devoted to AI governance, board presentations on AI risk, and oversight of the AI accountability chain is typically not tracked at all, despite representing a meaningful cost when calculated at fully-loaded executive rates.

> ***An incomplete cost model is not a financial analysis. It is a wish list with a spreadsheet attached.***

The result is systematic overstatement of AI's financial performance. When the business case shows a 2.8-times return, and the cost model used to generate it captures only the subscription fees and direct implementation costs, the return figure is not describing the performance of the AI investment. It is describing the performance of a subset of the AI investment against a subset of its costs. That subset may be impressive. The complete picture may be a different story entirely.

This matters for governance in a specific and practical way. If the board is approving AI investments based on incomplete cost models, the board is not exercising informed financial oversight. The Duty of Care from Chapter Three requires adequate information, and a cost model that excludes sixty percent of the actual costs is not adequate information. It is the financial equivalent of a balance sheet that omits a category of liabilities: technically organized, genuinely misleading.

The reason this gap persists is simpler than it appears. AI costs are distributed across organizational budgets, technology, legal, HR, finance, insurance, executive overhead, in ways that were established before AI existed as a distinct cost driver. No one function owns the complete picture. The CFO who manages capital allocation sees the technology budget line. The CFO who manages risk sees the compliance and insurance costs. The CFO who manages organizational efficiency sees the talent and oversight costs. In most organizations, those are the same person, and that person has never been given a framework that assembles all of those fragments into a single AI cost figure. This chapter provides the framework.

THE TCAIO MODEL: SEVEN CATEGORIES, ONE COMPLETE PICTURE

The Total Cost of AI Ownership model, TCAIO, assembles AI's full cost into seven categories. Each category answers a distinct financial question about what it actually costs to deploy, operate, and govern an AI system responsibly. Together, they produce the complete cost figure that a genuine ROI analysis requires, a board-level cost disclosure demands, and a fiduciary governance program is built on.

This framework is not an argument against AI investment. The CFO in this chapter's opening did not shut down her AI program when she saw the $1.2 million figure. She renegotiated three vendor contracts, consolidated two systems with overlapping functions, and implemented a more rigorous pre-approval process for new AI deployments. Within twelve months, the actual TCAIO had dropped to $820,000 against a revenue benefit that had increased as the retained systems were better governed and more reliably deployed. The program was cash-flow positive. The board had a number they could trust. Completing the cost model is not an obstacle to AI adoption. It is the foundation that makes AI adoption financially defensible.

CATEGORY 1, DIRECT LICENSING & SUBSCRIPTION COSTS

WHY IT BELONGS IN THE MODEL

This is the number that appears in the technology budget, and it is the number every AI business case is built around. It belongs in the TCAIO model not because it is hidden but because it is the anchor against which every other category is calibrated. Without it in the same model as the other six categories, the other six cannot be expressed as a proportion of the total.

WHAT'S INCLUDED

All vendor licensing fees, per-seat subscription costs, per-transaction API fees, platform access fees, and any usage-based charges that scale with the volume of AI activity. Include both current-year actuals and projected costs for systems under contract. For shadow AI systems identified through the Chapter Two detection process, include any costs authorized through the spend review.

MOST COMMON GAP

Failing to include API usage fees in this category because they appear on a cloud services bill rather than a vendor contract line item. API costs for AI services can exceed subscription costs for high-volume applications and should be captured here regardless of where they appear in the chart of accounts.

CATEGORY 2, INTEGRATION & MAINTENANCE INFRASTRUCTURE

WHY IT BELONGS IN THE MODEL

AI systems do not operate in isolation. They require data pipelines, storage capacity, processing infrastructure, and ongoing maintenance by technical staff. These costs are real AI costs that typically appear in the IT infrastructure budget as general technology investments rather than AI-specific line items. Without deliberate attribution, they are systematically excluded from AI cost analyses.

WHAT'S INCLUDED

Data storage and processing infrastructure directly attributable to AI workloads. IT staff time spent on AI system maintenance, updates, integration support, and incident response. Cloud compute costs associated with AI inference or training workloads. Third-party integration tools or middleware that AI systems depend on. Security tooling specifically deployed to protect AI-related data flows.

MOST COMMON GAP

Attributing none of the IT budget to AI costs on the grounds that the infrastructure 'would have been there anyway.' The incremental infrastructure cost driven by AI deployment belongs in the TCAIO model. If the organization would have needed a smaller data warehouse, fewer compute instances, or fewer IT support hours without the AI program, the difference is an AI cost.

CATEGORY 3, DATA GOVERNANCE & SECURITY OVERHEAD

WHY IT BELONGS IN THE MODEL

AI systems consume data at a scale and sensitivity level that most organizations' pre-AI data governance frameworks were not designed to manage. The additional data governance work that AI deployment creates, data quality programs, access controls, privacy reviews, data lineage documentation, security monitoring for AI-specific attack vectors, is a real cost that the AI program generates even when it is managed by a team that was already in place.

WHAT'S INCLUDED

Data quality and cleansing effort required to bring data to a standard the AI system can use reliably. Privacy and data protection reviews triggered by AI processing activities. Access control implementation and audit for AI system data inputs. Data lineage documentation required for governance and compliance. Security tooling and monitoring specifically addressing AI-related risk vectors, including model inversion attacks, data poisoning, and prompt injection for generative AI systems.

MOST COMMON GAP

Excluding data governance costs on the grounds that 'we should be doing this anyway.' That may be true. It does not change the fact that the AI system is the proximate cause of the incremental governance work. The cost belongs in the TCAIO model whether or not the organization agrees the work is independently justified.

CATEGORY 4, TALENT, GOVERNANCE, OVERSIGHT & RETRAINING

WHY IT BELONGS IN THE MODEL
This is the category most consistently excluded from AI cost models, and the one whose absence most reliably produces the greatest gap between reported and actual cost. Every AI system that operates at Tier 2 or Tier 3 requires ongoing human oversight, people whose job includes reviewing outputs, monitoring performance, managing exceptions, investigating anomalies, and maintaining the governance documentation the accountability chain requires. That oversight work is not free. It is performed by people with fully-loaded compensation costs. And for most organizations, it had been performed informally, without attribution to the AI program at all.
WHAT'S INCLUDED
Staff time devoted to AI output review, monitoring, and exception management, calculated at fully-loaded cost, not just base salary. Training and retraining costs as AI systems change and staff roles evolve. New hire costs for governance-specific roles if the AI program has created the need for dedicated AI oversight capability. Executive and board time devoted to AI governance functions, including reading governance reports, attending AI-related committee meetings, and conducting the oversight activities this book describes, calculated at fully-loaded executive rates.
MOST COMMON GAP
Excluding executive and board governance time from the model on the grounds that 'executives would be spending time on something regardless.' AI governance is not a zero-cost overhead activity, it displaces other uses of executive time. The opportunity cost is real. For organizations with multiple Tier 3 systems, the aggregate board and executive governance time can represent a material cost that deserves explicit measurement.

CATEGORY 5, REGULATORY & COMPLIANCE COST

WHY IT BELONGS IN THE MODEL
The legal and regulatory environment for AI is not static. It is expanding. Every AI system in a regulated domain creates ongoing compliance obligations: fair lending assessments for credit-related AI, HIPAA and privacy reviews for health-related AI, employment law assessments for hiring AI, financial regulation compliance for AI touching financial products or reporting. These costs did not exist before the AI deployment. They are a direct consequence of it.
WHAT'S INCLUDED
Legal review costs for AI-related vendor contract amendments, regulatory examination preparation, and compliance assessment work. External counsel fees for AI-specific regulatory guidance. Internal legal staff time attributed to AI governance and compliance activities. Costs associated with regulatory examination responses, corrective action plans, or consent order compliance if

applicable. Ongoing fair lending, bias, and disparate impact assessment costs for AI systems in regulated domains.

MOST COMMON GAP

Treating AI compliance costs as a general legal budget item rather than an AI program cost. If the legal budget grew after AI deployment, and the growth is attributable to AI-related work, the growth is an AI cost. Attribution requires a conversation with the General Counsel or Chief Legal Officer about what portion of the legal function's work is now AI-related. That conversation is also a governance act.

CATEGORY 6, INSURANCE & LIABILITY RESERVE

WHY IT BELONGS IN THE MODEL

Chapter Five established that Tier 3 systems require dedicated liability reserve consideration. That reserve is an AI cost. So are the insurance premium adjustments required to extend coverage to AI-related liability, the endorsements, riders, and policy modifications that bring AI-generated outputs within the scope of the organization's technology errors-and-omissions, professional liability, and general liability coverage. Most organizations have made some insurance adjustments for AI. Few have included those adjustments in the AI program's cost model.

WHAT'S INCLUDED

Annual premium cost of AI-specific insurance endorsements or riders. Any dedicated liability reserves established for Tier 3 AI systems, expressed as an annualized cost of capital for the reserve amount. Deductibles and self-insured retentions applicable to AI-related claims. Any D&O insurance premium changes attributable to the organization's AI governance posture.

MOST COMMON GAP

Treating this category as zero because 'we haven't had an AI claim.' The absence of a claim does not mean the absence of exposure, and the cost of the reserve and coverage is real whether or not it is ever drawn upon. This is the same logic that applies to any other contingent liability reserve: the cost of prudent risk management belongs in the cost model.

CATEGORY 7, OPPORTUNITY COST OF ORGANIZATIONAL ATTENTION

WHY IT BELONGS IN THE MODEL

This is the category that generates the most resistance in cost modeling conversations, and the one whose exclusion most dramatically understates AI's real cost in organizations running multiple Tier 2 and Tier 3 systems. Organizational attention is finite. The time that executives, board members, legal teams, compliance functions, and senior managers devote to AI governance is time that is not available for other organizational priorities. It is not free. It has a cost that is measurable, and in organizations where AI governance is consuming meaningful hours at senior levels, that cost is material.

> ***WHAT'S INCLUDED***
> *Fully-loaded cost of executive hours devoted to AI governance activities: reviewing governance reports, conducting oversight interviews, attending AI committee meetings, briefing the board on AI risk, managing vendor accountability conversations, and performing the oversight functions this book describes. Fully-loaded cost of board member hours devoted to AI governance: reading governance summaries, attending technology committee meetings focused on AI, reviewing incident reports, and conducting the board-level oversight the Duty of Care requires. The calculation is straightforward: hours multiplied by the fully-loaded hourly rate of the role.*
> ***MOST COMMON GAP***
> *Declining to include this category because the governance work 'should be done regardless.' The point of the opportunity cost calculation is not to argue against doing the governance work. It is to make visible what the governance work actually costs, so the organization can make informed decisions about how to staff it, prioritize it, and improve its efficiency over time. An organization that does not know what its AI governance costs cannot optimize it.*

Seven categories. One complete picture. Here is the one-page template that assembles them:

Exhibit 3: TCAIO Cost Model

EXHIBIT 3: TCAIO TOTAL COST OF AI OWNERSHIP MODEL				
#	**Cost Category**	**What to Include**	**System Name: [ ]**	**Annual Cost ($ or estimate)**
1	**Direct Licensing & Subscriptions**	Annual SaaS subscription, per-seat licenses, API usage fees, enterprise license costs. The number in the budget line.	$__________	
2	**Integration & Maintenance Infrastructure**	Cloud compute, storage, data pipeline maintenance, IT staff time allocated to this system, monitoring tools.	$__________	
3	**Data Governance & Security Overhead**	Data classification, access controls, privacy review, security audits, DPA or GDPR compliance work attributable to this system.	$__________	
4	**Talent: Governance, Oversight & Retraining**	FTE hours for governance oversight, output review, model retraining coordination, vendor management. Use fully loaded cost rate.	$__________	
5	**Regulatory & Compliance Cost**	Legal review hours, compliance assessments, regulatory filing preparation, external counsel attributable to this system.	$__________	
6	**Insurance & Liability Reserve**	Premium allocation for errors & omissions, cyber, and D&O coverage attributable to AI exposure. Pro-rate if portfolio-level.	$__________	
7	**Opportunity Cost of Attention**	Executive and board hours spent on governance, oversight, and reporting for this system. Fully loaded rate x hours/year.	$__________	

TOTAL TCAIO	$__________
Reported ROI (from original business case)	__________ x
Actual ROI = (Annual Benefit - TCAIO) / TCAIO	__________ x

Complete one template per AI system. Aggregate to portfolio level for board reporting. If actual ROI is below 1.0x, the system is cash-flow negative. Present to Executive Sponsor before next board cycle.

TCAIO ONE-PAGE CALCULATION TEMPLATE

Complete one row per AI system, aggregate across the portfolio for total organizational exposure

#	COST CATEGORY	ANNUAL ESTIMATE	NOTES / ASSUMPTIONS	% OF TOTAL
1	Direct licensing & subscriptions	$______	*Vendor contracts, per-seat fees, API usage*	___%
2	Integration & maintenance	$______	*IT staff time, infrastructure, updates*	___%
3	Data governance & security	$______	*Storage, access controls, audit tooling*	___%
4	Talent, governance & oversight	$______	*Staff time for review, training, ownership*	___%
5	Regulatory & compliance	$______	*Legal review, assessments, exam prep*	___%
6	Insurance & liability reserve	$______	*AI endorsements, dedicated reserves*	___%
7	Opportunity cost of attention	$______	*Executive & board time at fully-loaded rate*	___%
	TOTAL ANNUAL TCAIO	**$______**		**100%**

Reported ROI (from business case): **$______**

Actual ROI (Revenue benefit − TCAIO): **$______**

ROI Ratio (Actual ÷ Reported): ______×

Note: *Opportunity cost of attention (Category 7) is calculated as the fully-loaded hourly rate of each role multiplied by hours per year devoted to AI governance, oversight, and board reporting.*

> ***CFO NOTE***
> *Complete one TCAIO template per AI system, then aggregate across the portfolio. The aggregate figure is the number your board needs to see alongside the reported ROI. The ratio of actual ROI to reported ROI tells you whether your AI program is generating the financial performance your governance approved, or whether the business case was measuring a subset of a larger financial reality.*

THE NUMBER THAT CHANGED THE CONVERSATION

The CFO's first instinct was to find the error. I have seen this in every version of this conversation. Maybe the infrastructure costs were overstated. Maybe the executive governance time was being double-counted. Maybe the liability reserve should not be included because it had not been drawn upon. We went through each category together. The infrastructure costs held. The executive time calculation was, if anything, conservative, it used base salary rather than fully-loaded rates for the oversight staff. The liability reserve was real and was in the risk committee's minutes. The number was $1.2 million.

Once she accepted the number, she did something that separated her from many of the executives I have worked with on this problem. She did not try to hide it from her board. She brought it to the next audit committee meeting with a presentation that walked through the TCAIO model, explained each category, and showed the ROI calculation both ways: the 2.8-times return the board had approved, and the 0.8-times return against the actual cost.

The audit committee chair, a former investment banker with thirty years of capital markets experience, looked at the two figures for a long time. Then he said: "Every capital allocation decision we've made in this company for the last twenty years has been made on a fully-loaded cost basis. Why were we approving AI on a partial cost basis?"

Nobody had a good answer. Because nobody had asked the question before.

What followed was a governance upgrade that had nothing to do with technology and everything to do with financial discipline. The CFO commissioned a TCAIO analysis for every AI system in the register. The analysis took six weeks. The findings: three of the organization's nine AI systems were cash-flow positive on a TCAIO basis. Four were cash-flow negative but strategically important, meaning the board needed to make an explicit decision to fund them as strategic investments rather than cost-efficient operations. Two were cash-flow negative with no clear strategic justification, they had been approved on the basis of reported ROI figures that did not survive the full-cost analysis.

Those two systems were decommissioned. The resources freed by their decommissioning were redeployed to the governance infrastructure of the three cash-flow-positive systems, increasing their oversight intensity and, ultimately, their reliability. The CFO presented the completed TCAIO portfolio analysis to the board at the following quarterly meeting. The board approved a new AI investment approval standard: all future AI deployments would require a TCAIO estimate before the business case could be presented for approval.

The board had information it could trust. The CFO had a model she could update. The governance function had a financial foundation it could build on.

BUILDING YOUR COST MODEL THIS QUARTER

The TCAIO analysis does not need to be perfect to be useful. A directionally accurate cost model that captures all seven categories imprecisely is more valuable for governance purposes than a precise model that captures one category precisely and ignores the other six. Four actions that build the model.

1. **Start with your highest-cost Tier 3 system.**

Do not attempt to build a TCAIO model for your entire AI portfolio simultaneously. Start with the single AI system that, based on what you already know, is likely to have the highest total cost when all seven categories are included. This is almost always a Tier 3 system, because Tier 3 systems drive the most insurance, compliance, oversight, and governance cost. Build the TCAIO for that one system, using the template in this chapter. The exercise will take between half a day and a full day, depending on how distributed the cost data is across your organization. When it is complete, you will have a model and a methodology that can be applied to every other system in the register.

Start with the system that carries the highest governance consequence, not the highest sticker price. A Tier 3 underwriting model with a $40K licensing fee carries more TCAIO complexity than a Tier 1 analytics tool with a $400K enterprise license. Governance overhead, legal review time, and insurance exposure concentrate at the high-consequence tier, not the high-cost tier. Success looks like a completed TCAIO model for every Tier 3 system before the first Board AI Governance Summary is prepared.

2. **Convene a one-hour cost attribution meeting with Finance, Legal, and IT.**

Categories 2 through 6 of the TCAIO model require cost data that lives in budgets owned by functions other than the technology team. The most efficient way to gather this data is a single one-hour meeting with the key budget owners: Finance (for insurance and liability reserve data), Legal (for compliance and legal review cost data), and IT (for infrastructure and maintenance cost data). The meeting has a single agenda: for the AI system being modeled, what does each function estimate it is spending, in staff time and direct costs, that is, attributable to that system? The estimates do not need to be audited. They need to be directionally accurate and documented.

Run the meeting live, not by email. Email requests for cost data return incomplete, inconsistent responses weeks later. A one-hour live session with Finance, Legal, HR, and IT produces a rough TCAIO in real time, with the people who own the data making the attribution judgments together. Estimates are acceptable at this stage. The goal is order of magnitude, not audit precision. Success looks like a seven-category cost estimate you could defend in an executive conversation, not a spreadsheet requiring three follow-up rounds to complete.

3. **Calculate the ROI ratio and present it alongside the reported ROI.**

Once the TCAIO is complete for a given system, calculate the actual ROI: the revenue benefit or cost savings the system generates, minus the total TCAIO, divided by the total TCAIO. Compare this to the reported ROI from the original business case. The ratio of actual ROI to reported ROI is the number your board needs to see. A ratio above 1.0 means the system is performing better than the business case projected. A ratio between 0.5 and 1.0 means the system is performing but less efficiently than approved.

A ratio below 0.5 means the governance conversation about whether this system should continue in its current form is overdue.

Most common mistake: presenting the TCAIO figure without the ROI comparison, which turns the model into an accounting exercise rather than a decision-making tool. The TCAIO's governance value is its ability to show the board whether AI investments are performing as approved. Without the ROI ratio, you have a cost figure. With the ROI ratio, you have a capital allocation judgment. Present both. Success looks like a one-page summary for each analyzed system showing: reported ROI from business case, actual TCAIO, revenue benefit, actual ROI, and ROI ratio.

4. **Adopt the TCAIO as the standard for new AI investment approvals.**

The TCAIO model produces its greatest governance value not as a retrospective audit tool but as a prospective approval standard. Once you have completed the model for your current systems and presented the results to the board, propose adopting it as the required cost framework for all future AI investment approvals. Any AI deployment that requires board or executive approval should include a TCAIO estimate alongside the reported ROI projection. This single change closes the financial information gap that allows underperforming AI investments to be approved and protects the board's ability to exercise genuine financial oversight of the organization's AI portfolio.

Bring the motion to the meeting where you present the first TCAIO results. The CFO who presents the total cost of AI ownership model and then defers the approval standard to a future cycle has informed the board without equipping them to govern. The model and the standard belong together. One tells the board what AI actually costs. The other tells them the basis on which future investments will be evaluated. Success looks like a board that leaves the meeting having received both, and having approved a governance standard that every future AI business case must clear.

THE FINANCIAL FOUNDATION OF ACCOUNTABLE GOVERNANCE

The CFO whose board unanimously approved a $340,000 AI investment that turned out to cost $1.2 million was not practicing bad financial governance. She was practicing good financial governance with an incomplete cost model. The incompleteness was structural, not personal, the cost distribution across organizational budgets, the absence of a framework that assembled the fragments, and the inherited convention that AI costs meant technology budget costs. The TCAIO model is the framework that closes that structural gap.

The executive who builds a complete AI cost model is not the one who slows AI adoption. They are the one who makes AI adoption financially defensible, board-approvable, and sustainable past the first business case cycle. An organization that governs AI with a complete cost model makes better capital allocation decisions, retains its better-performing systems, decommissions the ones that do not perform, and presents its board with the financial information that genuine oversight requires.

The TCAIO model does not change what AI costs. It changes what executives know about what AI costs, and executives who know make different decisions: which systems to keep, which to decommission, which business cases to return to the team that wrote them. That discipline is the financial foundation of everything that follows. The next question is not about cost. It is about accountability: when something inside this portfolio goes wrong, and something will, whose name is on the obligation?

Governance Toolkit: Download live, editable versions of every template in this book:
brianzahn.gumroad.com/l/jgnqm

> ***AI is not free because you cannot see all its costs. The executive who builds the complete model is not the one who slows AI, they are the one who makes it last.***

BOARDROOM TAKEAWAY | *The Chain*

What a board member should be able to ask:

- For every Tier 2 and Tier 3 system, can we name the accountability owner at every link in the chain from vendor to board?
- Has each link in the accountability chain accepted their role in writing?

What a CFO or risk leader should have documented:

- A completed Six-Link Accountability Chain document for every Tier 2 and Tier 3 system
- Written acceptance of accountability from each named link in every chain

What evidence should exist:

- Chain documentation predating any incident: a record showing accountability was assigned before it was needed
- At least one tabletop exercise conducted for the highest-consequence Tier 3 system, with documented findings

CHAPTER SEVEN: PART II: THE FRAMEWORK

The Chain

Building Accountability From Vendor Contract to Board Room

When the AI system failed, nobody knew whose job it was to say so.

> **Illustrative Scenario:** *The following scenario is representative of documented patterns in executive AI governance. Names, organizations, and specific details are illustrative.*

The failure was not subtle. An AI-powered eligibility system at a mid-size benefits administrator had been miscategorizing claims for eleven weeks. Not occasionally. Systematically. A logic error introduced during a vendor-side model update had caused the system to apply the wrong eligibility criteria to one specific plan type, resulting in incorrect denial notices going out to approximately twelve thousand members. The denial notices referenced correct policy language. The underlying eligibility determination was wrong. Members who should have been approved were being denied, and because the denials were administratively clean, proper format, valid policy citations, correct signature blocks, most of them did not appeal. They simply accepted the denial and went without.

The error was discovered by a benefits coordinator in a regional office who noticed that she was receiving an unusually high volume of member complaint calls about a specific plan type. She flagged it to her supervisor. Her supervisor forwarded it to the operations manager. The operations manager sent an email to the technology team asking whether there had been any recent system changes. The technology team confirmed that a model update had been pushed in the relevant window. The operations manager escalated to the VP of Operations.

That process took nine days. By the time the VP of Operations asked the question that should have been asked on day one, who is the accountable executive for this system, and what is our incident response protocol, there was no clear answer to either part of the question. The vendor contract did not specify a notification obligation for model updates. The technology team had been monitoring model accuracy at the aggregate level and the aggregate accuracy had not degraded measurably. The VP of Operations knew the system existed. She did not know she owned its governance outcomes.

The answer to 'who is responsible for this' was excavated over four days of internal escalations, contract reviews, and organizational chart consultations. By the time it was settled, the system had been running in its failed state for eleven weeks and twelve thousand incorrect denial notices had been delivered. The organization's remediation, reprocessing twelve thousand claims, issuing corrected notices, managing regulatory notification, and addressing member relations, cost four times what a functioning accountability chain would have cost to build.

The accountability chain is not a governance nicety. It is the structure that converts an eleven-week failure into a two-day response. This chapter builds it.

WHEN EVERYONE IS RESPONSIBLE, NO ONE IS

The benefits administrator's eleven-week failure was not the product of negligence. Every person in the escalation chain did exactly what their role required them to do. The benefits coordinator flagged an anomaly. Her supervisor forwarded it. The operations manager investigated. The technology team confirmed. The VP of Operations escalated. Every one of those actions was correct and appropriate. The problem was not individual behavior. It was the absence of a pre-built accountability chain that told each of those people, in advance, who to call and how quickly.

That absence is the problem this chapter solves. And it is more widespread than most executives recognize, because accountability for AI is almost never absent entirely, it is almost always diffused. The technology team is accountable for the model. The business function is accountable for the process. The legal team is accountable for the contract. The compliance function is accountable for the regulatory posture. Each of those accountabilities is real. None of them is complete. And when something goes wrong, the gap between each function's accountability and the next function's accountability is where the eleven weeks live.

> ***Accountability without structure is intention. Accountability with a chain is governance. The gap between those two statements is where every AI failure gets its room to breathe.***

The reason most organizations have not built this structure is the same reason most governance structures are built reactively rather than proactively: the need for a pre-built accountability chain only becomes viscerally obvious after an incident that reveals its absence. Before the incident, the accountability diffusion is invisible. Everyone assumes that someone else's accountability covers the gap. After the incident, the gaps are mapped during the remediation process, expensively, reactively, and under conditions that make rigorous design impossible.

This chapter gives you the structure before the incident.

That is the entire point.

The cost of the absent accountability chain is not limited to incident response time. It shows up in every dimension of AI governance discussed in the prior chapters. The AI asset register is incomplete without named accountability owners at every link. The tier classification has no governance teeth without an escalation protocol that specifies who receives the finding and within what timeframe. The TCAIO model cannot capture the governance oversight cost without knowing which roles carry that oversight obligation. The accountability chain is not one governance structure among several. It is the connective tissue that makes the others functional.

THE SIX-LINK ACCOUNTABILITY CHAIN

The AI Accountability Chain assigns a named role at six points in the governance structure of every AI system. Each link has a defined responsibility, a defined escalation trigger, and a defined escalation path to the next link. When all six links are populated for a given system, the organization can answer three questions instantly in the event of a failure: who responds, who reports, and who owns the outcome. When any link is missing, the failure that exposes the gap will fill it, at the worst possible time, under the worst possible conditions.

The chain is a governance specification, not an org chart. A single person can hold multiple links for different systems. A single role can hold multiple links for the same system if the organization's structure supports it. What cannot be compressed is the function each link performs, the vendor layer, the technical layer, the business layer, the risk layer, the fiduciary layer, and the oversight layer are distinct governance functions even when the people performing them are the same people performing other organizational functions. Each link is defined below.

Exhibit 4: AI Accountability Chain

EXHIBIT 4: AI ACCOUNTABILITY CHAIN				
System Name: ____________ Tier: ___ Last Reviewed: ________				
Link	**Role**	**Responsibility**	**Named Individual / Title**	**Escalation Trigger**
1	**Vendor Accountability (Contract Layer)**	Vendor performance obligations, incident notification, model retraining disclosure, regulatory cooperation.	[Vendor contact name] [Account manager / legal] [Contract effective date]	Any output anomaly, model update, or regulatory inquiry involving this system
2	**Technical Owner (Model Layer)**	System configuration, output monitoring, override rate tracking, model version documentation.	[Name, Title] [Department] [Contact]	Override rate crosses threshold; vendor notification received; system output pattern change detected
3	**Business Owner (Decision Layer)**	Business outcome accountability, use case governance, decisions to expand, restrict, or decommission.	[Name, Title] [Department] [Contact]	Material output affecting customers, employees, or financial outcomes; escalation from Technical Owner
4	**Risk Officer (Governance Layer)**	Risk classification maintenance, governance policy compliance, incident log, board reporting input.	[Name, Title] [Contact]	Regulatory inquiry; incident declaration; tier reclassification required
5	**Executive Sponsor (Fiduciary Layer)**	Executive accountability for this system. Board-level reporting. Final decision authority on governance status changes.	[C-Suite Name, Title] [Contact]	Any incident at Tier 3; regulatory examination request; material output causing organizational harm
6	**Board Committee (Oversight Layer)**	Quarterly board visibility. Receives Board AI Governance Summary. Approves governance policy. Final oversight authority.	[Committee name] [Chair name]	Tier 3 system incident; regulatory enforcement action; material governance gap identified
Complete Links 1-6 for all Tier 3 systems. Links 1-5 for Tier 2 systems. Links 1-3 for Tier 1 systems. Each link holder must confirm their obligations in writing during the chain confirmation workshop.				

LINK 1, VENDOR ACCOUNTABILITY

The contract layer

The foundational question every AI accountability chain must answer before any internal link is built: what has the vendor committed to in writing? Vendor accountability covers three domains. Performance accountability: what specific performance standards does the contract obligate the vendor to meet, and what remedy exists when those standards are not met? Disclosure accountability: is the vendor required to notify your organization if the model is retrained, if a material performance issue is detected, or if a data breach involving your data occurs, and within what timeframe? Incident accountability: in the event of a system failure or harmful output, what is the vendor's documented obligation to cooperate with your organization's response, your regulatory response, and any required remediation?

ESCALATION PROTOCOL

Escalation trigger: any vendor notification of model change, performance degradation, data incident, or regulatory inquiry. Escalation path: to Link 2 (Technical Owner) within 24 hours; to Link 4 (Risk Officer) within 72 hours if the trigger meets the material threshold defined in the governance policy.

LINK 2, TECHNICAL OWNER

The model layer

The Technical Owner is the internal role responsible for the AI system's operational performance: monitoring its outputs, detecting anomalies, managing vendor interactions at the technical level, and maintaining the system documentation that governance reviews require. This is typically an engineering lead, a data science manager, or a senior IT professional with specific accountability for this system's technical health. The Technical Owner is not the accountable owner for governance purposes, that is, Link 5. They are the accountable owner for operational performance: they know what the model is doing, they are the first to know when something changes, and they are the person whose job includes raising a flag when the output pattern shifts in ways that warrant governance attention.

ESCALATION PROTOCOL

Escalation trigger: any anomaly in output distribution, accuracy degradation, override rate movement beyond the threshold established in the tier classification, vendor notification of model change, or operational incident affecting system availability or data integrity. Escalation path: to Link 3 (Business Owner) and Link 4 (Risk Officer) simultaneously within 24 hours of detection.

LINK 3, BUSINESS OWNER

The decision layer

The Business Owner is the functional leader whose team uses the AI system's outputs to make business decisions. They are accountable for ensuring that the system's outputs are being used in a manner consistent with its approved authority designation, that an advisory system is being used advisorily, that override protocols are being followed for assistive systems, and that autonomous systems are operating within the parameters that governance has approved. The Business Owner is typically a VP or director whose function relies on the system: the VP of Lending for a credit model, the Chief People Officer or HR director for a hiring tool, the VP of Operations for a logistics optimization system. They are not a governance role, they are the production role that has the clearest, most current visibility into how the system's outputs are being used in practice.

ESCALATION PROTOCOL

Escalation trigger: any pattern of outputs that appears inconsistent with the system's approved authority, any operational decision to override the system at unusual rates, any customer or employee complaint that appears linked to the system's outputs, or any business outcome that suggests the system may be performing differently than expected. Escalation path: to Link 4 (Risk Officer) within 48 hours; to Link 5 (Executive Sponsor) if the issue has been present for more than 72 hours without resolution.

LINK 4, RISK OFFICER

The governance layer

The Risk Officer is the organizational role responsible for maintaining the AI governance program: updating the AI asset register, conducting tier reclassification reviews, commissioning and reviewing output audits, assessing bias and fairness indicators, and producing the governance documentation that board reporting requires. This role may be a Chief Risk Officer, a Chief Compliance Officer, a dedicated AI governance lead, or a senior risk management professional with explicit AI accountability in their job description. The Risk Officer is the organizational memory of the AI governance program, the person who knows the history of each system's governance record, who has reviewed the prior audit results, and who can tell the Executive Sponsor and the board committee not just what is happening today but how it compares to what was happening six months ago.

ESCALATION PROTOCOL

Escalation trigger: any finding from a tier review, output audit, or bias assessment that meets the material threshold defined in the governance policy; any escalation from Link 2 or Link 3 that has not been resolved at the operational level within 72 hours; any regulatory inquiry, examination notice, or legal hold that involves an AI system. Escalation path: to Link 5 (Executive Sponsor) immediately upon meeting any escalation trigger; to the board committee (Link 6) if the issue meets the board-notification threshold defined in the governance charter.

LINK 5, EXECUTIVE SPONSOR

The fiduciary layer

The Executive Sponsor is the C-suite leader who carries board-level accountability for the AI system's governance. Not organizational responsibility for the system's business outcomes, that is, Link 3. Not technical responsibility for the system's performance, that is, Link 2. Governance accountability: the obligation, under the three fiduciary duties established in Chapter Three, to be adequately informed about the system's risk posture, to act in the organization's genuine interest in managing that risk, and to be able to demonstrate, to the board, to a regulator, to a court, that responsible oversight existed. The Executive Sponsor is typically the CFO for financial AI systems, the Chief People Officer for employment AI, the Chief Medical Officer for clinical AI, or the CEO for AI systems whose impact spans the organization broadly. The Executive Sponsor's name is the name that appears in the AI asset register's accountable owner column for Tier 3 systems.

ESCALATION PROTOCOL

Escalation trigger: any board notification threshold event defined in the governance charter; any regulatory inquiry or examination notice; any litigation hold or discovery request involving an AI system; any incident that has not been resolved at the Risk Officer level within the timeframe specified in the governance policy; any finding from an output audit that requires board disclosure under the organization's governance charter. Escalation path: to Link 6 (Board Committee) with written notice within the timeframe established in the governance charter, typically 48 hours for material findings and immediately for regulatory or litigation triggers.

LINK 6, BOARD COMMITTEE

The oversight layer

The Board Committee is the governance body that receives the AI governance brief, reviews material findings, challenges the adequacy of the governance structure, and exercises the board-level oversight that the Duty of Care requires. This is typically the risk committee, the audit committee, or a dedicated technology governance committee depending on the organization's board structure. The board committee's role in the accountability chain is not operational, they do not manage the AI system, review its outputs, or direct its governance. Their role is to ensure that the governance structure is adequate to the risk, that material issues are being surfaced and addressed, and that the organization's AI governance posture is consistent with its fiduciary obligations. They exercise that role by receiving the governance brief from Chapter Eight, asking the questions the brief is designed to enable, and creating a board record that demonstrates active oversight.

ESCALATION PROTOCOL

Standing obligation: receive the AI governance brief at every reporting cycle. Specific escalation: any event that triggers an Executive Sponsor notification is automatically included in the next board committee communication, with a status update provided outside the standard cycle if the event has material implications for the organization's risk posture. The board committee's receipt of the notification is itself a governance act, it should be documented in the committee's minutes.

> ***The Chain in Practice***
> *Every AI system in your AI asset register. Tier 2 and above, requires a populated accountability chain. For Tier 1 systems, Links 1 through 3 are sufficient. For Tier 2 systems, Links 1 through 5 are required. For Tier 3 systems, all six links are required, and the Executive Sponsor must be a named C-suite role, not a director or VP. The chain is documented in the AI asset register alongside the five columns from Chapter Four, as a seventh column titled 'Accountability Chain.'*

WHAT THE CHAIN FOUND

> **Illustrative Scenario:** *The following scenario is representative of documented patterns in executive AI governance. Names, organizations, and specific details are illustrative.*

A logistics company with approximately $900 million in annual revenue had been using an AI-powered routing optimization system for sixteen months when a significant service failure occurred. The system, which managed last-mile delivery routing across a regional network, began producing routes that were systematically inefficient in one geographic cluster. Deliveries that should have taken four hours were taking seven. Driver overtime costs increased sharply in that cluster. Customer satisfaction scores for on-time delivery dropped by twenty-two percentage points in six weeks.

The operations team noticed the overtime cost spike first. They attributed it to seasonal volume. When the customer satisfaction data surfaced, the VP of Operations commissioned an investigation. The investigation took eleven days to determine that the routing system was the cause, because no one had been monitoring the system's output quality at the cluster level, only at the aggregate network level, where the inefficiency was diluted.

The discovery of the cause triggered a secondary crisis: what had caused the routing system to fail? The technology team examined the model. They found that the vendor had silently retrained the model six weeks prior using updated map data. The updated map data contained an error in the road network for the affected geographic cluster, a new distribution center had been mapped incorrectly, causing the model to route deliveries through a network configuration that did not match physical reality. The vendor had not notified the logistics company of the retraining. The vendor contract did not require notification.

The VP of Operations called the vendor. The vendor acknowledged the issue and offered to retrain the model with corrected data. The call took forty minutes. The fix took seventy-two hours. The service failure had been running for six weeks.

In the aftermath of the remediation, the VP of Operations, who had never previously engaged with AI governance as a formal discipline, did something unusual. She called the organization's General Counsel and asked a single question: if this happens again, what structure do we need to have in place so that we find out in days instead of weeks, we know who calls the vendor, and we know who tells the board?

The General Counsel said: "You need an accountability chain. Let me show you what that looks like."

The chain they built over the following three weeks was straightforward in retrospect and would have been straightforward in prospect, had anyone thought to build it before the failure. Link 1 specified that the vendor contract required written notification of any model retraining within forty-eight hours, with a description of the data used and the expected behavioral change. Link 2 assigned the routing system technical owner to a senior engineer whose role explicitly included monitoring cluster-level performance metrics and escalating anomalies to Link 3 within twenty-four hours of detection. Link 3 assigned the routing system business owner to the VP of Operations herself, whose role included reviewing the Link 2 escalation report and making the decision to escalate further. Links 4 through 6 followed the standard escalation protocol.

The VP of Operations then renegotiated the vendor contract to include the notification requirement. The vendor pushed back. She held the line, citing the six-week service failure as evidence of the commercial necessity. The vendor agreed. The amendment was signed.

Eight months later, the vendor retrained the model again. This time, the notification arrived within twenty-four hours. The technical owner reviewed the retraining parameters, identified no material change to the routing logic, and documented the review in the governance record. The VP of Operations received the Link 2 summary. No escalation was required. The entire governance cycle took seventy-two hours and produced a documented record that the organization had exercised oversight of a vendor-side model change. When the organization's next insurance renewal came up, their broker asked about AI governance documentation. The VP of Operations provided the accountability chain and the documented vendor notification cycle. The broker noted that the governance record was among the most complete she had seen from an organization of that size, and recommended an improved premium structure accordingly.

The chain did not just tell the organization who was responsible when something went wrong. It defined what responsible looked like when things were going right.

BUILDING THE CHAIN BEFORE THE INCIDENT

The accountability chain does not require a governance consultant, a lengthy design process, or a new organizational structure. It requires a decision about who fills each link for each system, documented in writing, and embedded in the AI asset register before the next incident surfaces the absence. Four actions that build it.

1. Populate the chain for every Tier 3 system in your register this week.

Your Tier 3 systems are the systems whose failures will be the most consequential, the most visible, and the most likely to trigger regulatory or litigation exposure. They are also the systems most likely to be missing a complete accountability chain, because their governance complexity has discouraged the structured assignment work that less complex systems allow. Start there. For each Tier 3 system, open the AI asset register and add a seventh column. Accountability Chain, with six rows, one per link. Fill in the named role for each link. If a link cannot be filled because the appropriate role does not exist, or because there is genuine organizational disagreement about who should hold it, document the gap explicitly. An identified gap is a governance problem you can solve. An unidentified gap is a governance problem that will surface on its own schedule.

Name a person, not a function. "The risk management function" is not a Link 4 assignment. "Sarah Chen, Chief Risk Officer" is. The difference is the difference between accountability that can be exercised and accountability that can be disclaimed. When a regulator asks who is responsible for governing a specific AI system, a function does not appear in the examination room. A person does. Success looks like an accountability chain where every link holds a person's name, title, and direct contact information.

2. **Audit Link 1 for every Tier 2 and Tier 3 system: read the vendor contract.**

The vendor contract is the accountability chain's foundation, and it is the link most organizations have never examined through a governance lens. Pull the current executed contract for every Tier 2 and Tier 3 AI system. Read it, or have your General Counsel read it, for three specific provisions: the performance accountability clause (what does the vendor owe you when the model underperforms?), the retraining notification obligation (is the vendor required to tell you when the model changes?), and the incident cooperation clause (is the vendor obligated to cooperate with your regulatory and legal response if their system causes harm?). For each contract, document what exists and what is absent. The absent provisions are contract amendment priorities for the vendor negotiations in Chapter Twelve.

Read the contract before the meeting, not during it. General indemnification language does not cover these specific governance needs. The five provisions in Chapter Twelve are specific because general clauses produce general protection: legally real, practically inadequate. The Link 1 audit is a contract review exercise, not a relationship review. Success looks like a written summary, for each Tier 3 vendor, of which of the five provisions are present in the executed contract and which require amendment.

3. **Conduct a two-hour accountability chain workshop for each Tier 3 system.**

Once the chain is drafted on paper, convene a two-hour working session with the proposed Link holders for each Tier 3 system: the Technical Owner, the Business Owner, the Risk Officer, and the Executive Sponsor. The session's agenda is three questions. First: does everyone in this room know that they hold the link assigned to them, and do they accept the escalation obligations that link requires? Second: are the escalation triggers specific enough that each Link holder will know without ambiguity when they have been triggered? Third: has the chain ever been tested, through a tabletop exercise or simulated incident, to verify that it functions as designed? An untested chain is a design document. A tested chain is a governance structure.

Most common mistake: skipping the acceptance step and assuming that assigned roles are accepted roles. The Technical Owner, Business Owner, and Risk Officer for a given AI system may not know that they hold those governance designations, may not have received any briefing on the escalation obligations those designations carry, and may have competing priorities that make the governance obligations unworkable without organizational support. The workshop surfaces these issues before an incident, not during one. Success looks like three clear yes answers to the three workshop questions, documented in the meeting minutes.

4. **Run a tabletop exercise for your most consequential Tier 3 system.**

A tabletop exercise is a structured simulation of an AI incident, conducted with the full accountability chain in the room, to test whether the chain functions as designed under realistic conditions. The scenario does not need to be elaborate. Describe a realistic failure mode for the system, a systematic output error, a vendor retraining notification, a regulatory inquiry, a media report, and walk through each link's response. Who is notified first? What is the escalation timeline? What documentation is produced at each link? Where does the decision authority sit? What does the board communication look like? The exercise typically takes ninety minutes and surfaces between three and seven procedural gaps that the chain design did not anticipate.

Run the tabletop as a stress test, not a rehearsal. A tabletop exercise where everyone agrees on the next step is not testing the accountability chain. It is confirming the org chart. The exercise should surface the gaps: the step where two people believe the other is responsible, the escalation that has no defined timeline, the board notification trigger no one has agreed on. Those gaps, surfaced in the tabletop, are closed before the incident. Success looks like a tabletop that produces a list of governance corrections, because that list is the point of the exercise.

THE CHAIN THAT WAS ALREADY THERE

Every governance structure in this book has been built on a foundation that the accountability chain connects. The AI asset register identifies the systems. The tier model classifies their risk. The TCAIO model quantifies their cost. The accountability chain names who is responsible for every layer of every system's governance, from the vendor contract that defines what the vendor owes, to the board committee that exercises the oversight the Duty of Care requires. Without the chain, the other structures are instruments without hands. With it, every governance act has a named owner and every failure has a pre-specified response.

The logistics company's VP of Operations spent eleven weeks learning what the chain would have cost three weeks to build. The benefits administrator spent four months learning what the chain would have cost two weeks to build. In both cases, the cost of the absence was not a governance cost. It was an operational cost, a regulatory cost, a reputational cost, and in the benefits administrator's case, a human cost, twelve thousand people who went without benefits they were entitled to because the system's failure had no named owner and no pre-built response.

Accountability without structure is intention, and every organization has good intentions about AI governance. Accountability with a chain is governance, and far fewer organizations have built it. The distance between those two things is the distance this chapter closes.

You now have the inventory, the risk classification, the cost model, and the accountability structure. All of it exists. None of it is visible to your board. The question that determines whether you are actually governing AI, or simply intending to govern it, is this: can the people responsible for organizational oversight see what you have built? Chapter Eight builds the document that answers that question.

> ***When a failure occurs, the chain tells you exactly who responds, who reports, and who owns the outcome. Build it before the failure. Own it before the question is asked.***

BOARDROOM TAKEAWAY | *The Governance Layer*

What a board member should be able to ask:

- Has our board received a governance summary that addresses risk, accountability, and oversight rather than investment and capability?
- Could any director on our board, reading our most recent AI briefing, describe our governance posture to an examiner?

What a CFO or risk leader should have documented:

- A completed Board AI Governance Summary in the six-section format
- A confirmed quarterly cadence for the governance summary as a standing board committee agenda item

What evidence should exist:

- The board committee's documented receipt of the first governance summary
- A clear distinction, in board records, between AI capability briefings and AI governance briefings

CHAPTER EIGHT: PART II: THE FRAMEWORK

The Governance Layer

What Boards Actually Need to See

Ask the most experienced board member in your organization this question: 'When did you last receive a document that told you, on a single page, what AI systems your organization is running, which carry the most risk, what they cost in total, and who is accountable when one of them fails?'

In most organizations, the answer is never. Not because the board does not want that document. Because no one has built it. What boards receive instead is one of two things, each a failure mode in its own way.

> **Illustrative Scenario:** *The following scenario is representative of documented patterns in executive AI governance. Names, organizations, and specific details are illustrative.*

The board member had been on the risk committee for six years. She had reviewed hundreds of management reports. She knew how to read a financial statement, interrogate a capital allocation proposal, and press an executive team on assumptions she found optimistic. When the AI governance briefing landed in her board packet for the first time, she read it twice and then set it down.

It was twenty-three pages. It began with a section on transformer architecture. It included a discussion of model drift that used the phrase "epistemological uncertainty in stochastic inference environments." Page fourteen contained a chart with seventeen variables and no legend. The executive summary, which was eight paragraphs long, concluded with the observation that the organization's AI program was "operating within acceptable performance parameters relative to deployment-context benchmarks."

She had no idea what the document was asking her to do. She was not sure it was asking her to do anything. She approved it in the committee meeting because the CTO presented it confidently, nobody else asked questions, and the agenda had three more items after it. The committee approved it four to zero in six minutes.

After the meeting she pulled the Chief Risk Officer aside. She said: "I want to support our AI governance program. I don't know how to ask good questions about it. I don't know what I should be worried about. I don't know what this document is telling me. Can you give me something that tells me what I need to know to do my job on this committee?"

The Chief Risk Officer went back to her office and spent two weeks thinking about that question. The document she produced, a single-page governance summary with six sections, each written in plain language, each designed to enable a specific board oversight question, is the template at the center of this chapter. The board member who had asked for it told the CRO afterward that it was the first AI governance document she had ever been able to read without a translator. She has asked for it at every board meeting since.

This chapter builds that document.

THE REPORT THAT CANNOT BE INTERROGATED

Most board-level AI reporting exists on a spectrum between two failure modes. At one end is the technology briefing: a detailed, technically sophisticated document prepared by the technology function that demonstrates capability, describes architecture, and reports performance metrics against technical benchmarks. It is accurate. It is thorough. It is written for an audience that does not sit on most boards, and it produces board meetings in which the technology team presents and the board nods.

At the other end is the governance non-report: a brief assurance statement, typically embedded in a broader risk or audit committee report, noting that AI systems are "operating within expected parameters" or that the organization's AI governance program is "consistent with industry standards." These statements are not lies. They are also not governance. They give a board member no basis for a question, no basis for a concern, and no basis for a finding that anything requires attention. They produce board approval by default, because there is nothing specific enough to approve or challenge.

> ***A board that cannot interrogate AI governance is not exercising AI governance. It is providing a signature.***

The cost of both failure modes is the same: the board is not exercising the oversight the Duty of Care requires. A board that approves a twenty-three-page technology briefing it cannot interrogate is not performing AI governance. A board that accepts a two-sentence assurance statement it has no basis to challenge is not performing AI governance. Both boards are providing fiduciary cover without fiduciary oversight, and examiners, plaintiffs, and institutional investors are learning to tell the difference.

The reason this gap persists is a mismatch between the expertise of the people who prepare AI governance reporting and the expertise of the people who need to consume it. The technology team knows what information exists and how to present it to people who share their technical background. Board members know what governance questions to ask about domains they understand and have governing frameworks for, financial performance, legal exposure, strategic risk. Nobody has built the bridge between those two bodies of expertise: a reporting format that translates technical governance information into the language of fiduciary oversight.

The Board AI Governance Summary is that bridge. It does not simplify what is happening with AI. It translates what is happening into the terms that governance oversight requires: what do we have, what does it risk, what has gone wrong, what does it cost, who is accountable, and what is coming from regulators. Those six questions are not technical questions. They are governance questions. A board member who can answer all six, from a single-page document, is exercising the informed oversight the Business Judgment Rule requires.

THE BOARD AI GOVERNANCE SUMMARY: SIX SECTIONS, ONE PAGE

The Board AI Governance Summary condenses everything built in the prior four chapters, the AI asset register, the tier model, the TCAIO model, and the accountability chain, into a single-page document structured around the six questions a board member needs to answer to exercise meaningful AI oversight. Each section is written in plain language. Each section enables at least one specific board question. Each section contains a red flag signal that tells the board when something requires more than routine oversight attention.

The document is one page by design, not by constraint. Board governance documents that require extended reading before a question can be formed are governance documents that will not be interrogated. The one-page discipline forces the executive preparing the summary to make choices about what is governance-essential information versus what is operationally interesting but governance-peripheral. If it does not enable a board governance question, it does not belong on the page. Detailed supporting materials for any section can be appended, but the summary itself must stand alone as a complete governance picture.

SECTION 1, AI ASSET SUMMARY

PURPOSE

Give the board a complete inventory of what the organization is running, organized by governance significance, so they can ask whether the portfolio has grown appropriately and whether all systems have the governance coverage they require.

WHAT TO INCLUDE

Total number of AI systems currently active, broken down by tier. Number of systems added and decommissioned since last report. Confirmation that every system has a named accountable owner, or explicit identification of any system that does not. One sentence on any material change to the portfolio composition.

THE BOARD QUESTION THIS ENABLES

"How many autonomous AI systems are we running, and is every one of them covered by the governance structure we approved?"

RED FLAG SIGNAL

Any Tier 3 system without a named executive-level accountable owner. Any system added this period without a completed TCAIO analysis. Any system that has been operating without a tier classification.

SECTION 2, RISK POSTURE

PURPOSE

Give the board a current picture of the organization's AI risk distribution and flag any changes that require attention, so they can assess whether the risk profile is consistent with the organization's risk appetite and governance capacity.

WHAT TO INCLUDE

Current tier distribution. Any reclassifications since last report, with the reason for the change. Override rate flags for any system whose override rate has crossed the reclassification threshold. A one-sentence risk posture assessment comparing current status to last report: improving, stable, or elevated, with a single-sentence rationale. Scheduled tier reviews in the next reporting period.

THE BOARD QUESTION THIS ENABLES

"Has our risk profile changed since last quarter, and is there anything in the current tier distribution that the committee should be specifically concerned about?"

RED FLAG SIGNAL

Any Tier 1 system whose override rate has fallen below twenty percent without a reclassification review. Any new Tier 3 system deployed since last report. Any system whose risk posture assessment is 'elevated' for two consecutive reporting cycles.

SECTION 3, INCIDENT LOG

PURPOSE

Give the board a complete and honest account of every AI-related incident in the reporting period, including how it was detected, how it was resolved, and whether any items remain outstanding, so they can verify that the accountability chain is functioning and that incident response is adequate.

WHAT TO INCLUDE

Number of incidents this period and year-to-date. For each incident: a plain-language description of what happened, which system was involved, when it was detected, how it was detected (monitoring, escalation, complaint, external report), how it was resolved, and whether any member, customer, or employee harm occurred. Status of all open items from prior period incidents. Any incident that was escalated to the Executive Sponsor level or above.

THE BOARD QUESTION THIS ENABLES

"Were there any AI incidents this quarter, and for each one, how was it detected, how long did the response take, and is there anything unresolved that requires board attention?"

RED FLAG SIGNAL

Any incident that was detected by an external party (regulator, customer, media) before the internal accountability chain detected it. Any incident where the response time exceeded the escalation protocol timeframes. Any incident where harm to a third party occurred and has not been fully remediated.

SECTION 4, COST SUMMARY (TCAIO)

PURPOSE

Give the board a complete picture of what AI is actually costing the organization across all seven TCAIO categories, compared to what was approved, so they can assess whether the AI portfolio is performing as the business cases projected and whether any cost variance requires a governance decision.

WHAT TO INCLUDE

Total TCAIO for the current period, annualized, against the approved budget. Variance and its driver, explained in one sentence. The largest cost category and its percentage of total. Subscription costs as a percentage of total TCAIO, a ratio that tells the board how much of the true cost is visible in the technology budget versus distributed across other organizational budgets. Actual ROI ratio: the ratio of actual ROI (calculated against TCAIO) to reported ROI (calculated against subscription costs only).

THE BOARD QUESTION THIS ENABLES
"Is our AI portfolio generating the return we approved, and are there any cost variances that require the committee to make a decision or ask management for a remediation plan?"
RED FLAG SIGNAL
Any system whose actual ROI ratio has fallen below 0.5 for two consecutive reporting cycles without a board decision about its strategic justification. Any cost category variance exceeding twenty percent of budget without a documented explanation. Total TCAIO growing faster than the AI portfolio without a corresponding increase in documented business benefit.

SECTION 5, ACCOUNTABILITY STATUS

PURPOSE
Give the board a current status of the accountability chain for every Tier 2 and Tier 3 system, so they can verify that every AI system with significant governance obligations has a named human being accountable at every required link.
WHAT TO INCLUDE
Chain coverage status for Tier 3 systems: percentage of links named and confirmed for each system. Chain coverage status for Tier 2 systems: percentage of Links 1 through 5 named and confirmed. Any chain gaps, with the system name, the missing link, and the target date for resolution. Vendor contract audits completed this period and any amendments executed. Tabletop exercises completed and scheduled.
THE BOARD QUESTION THIS ENABLES
"Does every high-risk AI system in our portfolio have a named accountable owner at every governance link, and are there any gaps that management needs to resolve before next quarter?"
RED FLAG SIGNAL
Any Tier 3 system with an incomplete accountability chain for more than one reporting cycle. Any Link 5 (Executive Sponsor) vacancy that has not been filled within thirty days. Any Tier 3 system that has not had a tabletop exercise within the prior twelve months.

SECTION 6, REGULATORY HORIZON

PURPOSE
Give the board advance visibility into the regulatory requirements that are approaching, so they can verify that the organization is preparing adequately and so they can make informed decisions when regulatory developments require a board-level response.
WHAT TO INCLUDE
Active regulatory developments with a direct implication for the organization's AI governance posture, summarized in one to two sentences each. Preparation status for each: not started, in progress, or complete. Any development that is likely to

> *require a board decision or formal board approval within the next two reporting cycles. Any active regulatory examination, inquiry, or enforcement matter involving AI systems.*
> ***THE BOARD QUESTION THIS ENABLES***
> *"Are there regulatory developments on the horizon that this committee needs to understand, and is management's preparation adequate given our current governance posture?"*
> ***RED FLAG SIGNAL***
> *Any regulatory development with a compliance deadline within the next two reporting cycles that is still at 'not started' status. Any active regulatory examination involving an AI system that has not been disclosed to the full board. Any state or federal enforcement action in the organization's industry that signals a governance standard the organization's current posture does not meet.*

Six sections. One page. Here is the complete template, populated with illustrative data:

Exhibit 5: Board AI Governance Summary

EXHIBIT 5: BOARD AI GOVERNANCE SUMMARY			
[Organization Name] | Board AI Governance Summary | Quarter Ending [Q/YYYY] | Prepared by: [Executive Sponsor]			
SECTION 1 AI Asset Summary	Total AI systems: 14 Tier 3 (Autonomous): 2 Tier 2 (Assistive): 5 Tier 1 (Advisory): 7 New this quarter: 1	Changes from prior quarter: + 1 new Tier 2 deployment (HR screening) 0 decommissioned 0 reclassified	Tier 3 systems named: 1. Automated credit scoring (CreditScore AI) 2. Real-time fraud detection (SecureAI) Both systems have confirmed accountability chains.
SECTION 2 Risk Posture	Highest-risk system: CreditScore AI (Tier 2) Override rate: 18% Status: Within governance threshold	Monitoring status: All Tier 3 systems: current All Tier 2 systems: current Override rate reviews: completed Q3	Risk flags: HR Screening Tool upgrade planned Q4. Pre-deployment governance review scheduled. No systems currently outside governance threshold.
SECTION 3 Incident Log	Q3 incidents: 1 Severity: Low System: Meeting Summarizer (Tier 1) Nature: Transcript error causing summary inaccuracy	Resolution: Identified Day 2. Corrected summary distributed Day 3. No external impact. Vendor notified. Root cause: audio quality threshold.	Governance action taken: Added audio quality check to Technical Owner review protocol. No tier reclassification required. Documented in governance record.
SECTION 4 Cost Summary (TCAIO)	Total AI portfolio cost (TCAIO): $2.4M annually Reported budget line: $890K Governance overhead: $1.51M	Portfolio ROI (actual): 1.8x (positive) Prior quarter: 1.6x Improvement: Decommissioned 2 cash-negative systems in Q2	Systems with negative actual ROI: 0 Systems under ROI review: 1 (document processing tool, Q4 review scheduled)
SECTION 5 Accountability Status	Accountability chains: Tier 3: 2/2 complete and confirmed Tier 2: 5/5 complete Tier 1: Accountable owners named for all 7	Policy compliance: AI Governance Policy adopted: Q1 20XX Last reviewed: Q2 20XX Next annual review: Q1 20XX+1	Open governance actions: 1. Vendor contract amendment (Link 1, Tier 2 HR tool): in negotiation, target Q4 close 2. Tabletop exercise (Tier 3 SecureAI): scheduled Q4

SECTION 6 Regulatory Horizon	Active regulatory requirements: - ECOA / fair lending: compliant - OCC model risk: compliant - State biometric law: not applicable	Upcoming requirements: - Colorado AI Act: effective Feb 2026 Status: Gap assessment completed, remediation in progress - EU AI Act: not applicable (US-only operations)	Regulatory readiness posture: PREPARED for current requirements. IN PROGRESS on Colorado AI Act. No enforcement actions or regulatory inquiries pending.
This document is a board governance record. Prepare quarterly. Retain per document retention policy. Distribute to board risk or audit committee as a standalone agenda item, not embedded in the technology briefing.			

BOARD AI GOVERNANCE SUMMARY

[Organization Name]. Quarter Ending [Date]. Prepared by: [CFO / Executive Sponsor]

This document is a board governance record. Retain per document retention policy.

Section 1, AI Asset Summary

Total AI systems: 14 (1 added this quarter)
Tier 1 (Advisory): 7 systems
Tier 2 (Assistive): 5 systems
Tier 3 (Autonomous): 2 systems
Systems with named owner: 14 of 14, no gaps
New deployments this quarter: 1 (Tier 2, HR screening tool)
Decommissioned this quarter: 0

Section 2. Risk Posture

Tier 3 systems: LoanSense AI, DemandIQ, no change in tier
Reclassifications: None this quarter
Override rate flags: HR screening tool (new), first review in 90 days
Audit findings: 1 minor finding, resolved (see Section 3)
Risk posture vs. last quarter: Stable, no new Tier 3 deployments
Next scheduled tier reviews: LoanSense AI, Q3; all others Q4

Section 3. Incident Log

Incidents this quarter: 1
Description: LoanSense AI output anomaly detected in Week 6. Override rate increased to 18% (above 15% flag threshold). Technical Owner notified Risk Officer per protocol.
Resolution: Vendor confirmed no model change. Anomaly traced to upstream data feed issue. Resolved in 72 hours. No member harm documented.
Documentation status: Full incident record filed. Governance record updated.
Outstanding items: None, incident closed.
Incidents YTD: 2, both closed

Section 4. Cost Summary (TCAIO)

Total TCAIO (all systems): $1.04M annualized
Budget: $980K, 4% over plan
Variance driver: Legal review costs for new HR tool (Category 5). One-time. Not expected to recur.
Largest cost category: Talent, governance & oversight (31% of total)
Subscription costs as % of total: 28%, below industry avg.
TCAIO vs. reported ROI: Actual ROI ratio: 1.4×, above business case projection

Section 5. Accountability Status

Section 6. Regulatory Horizon

Chain coverage. Tier 3: 100%, all six links named and confirmed

Chain coverage. Tier 2: 100%. Links 1–5 named and confirmed

Chain coverage. Tier 1: 100%. Links 1–3 named and confirmed

Changes this quarter: HR screening tool chain populated at deployment

Vendor contract audits: 1 amendment executed (LoanSense AI, retraining notification added)

Tabletop exercise: DemandIQ, completed Q1. LoanSense AI scheduled Q3.

EU AI Act: High-risk classification review in progress. Legal engaged. Estimated completion: Q3.

CFPB AI guidance: Monitoring. No new enforcement guidance since last report.

State AI legislation: 3 states with pending AI bills affecting credit decisions. Tracking via outside counsel.

NIST AI RMF: Gap assessment completed Q4 prior year. 4 of 7 recommendations implemented.

Next regulatory review: Q3 board report will include full regulatory readiness matrix.

Action required by board: None this quarter. Q3 may require EU AI Act classification approval.

Executive Sponsor Certification: *I certify that this summary is accurate to the best of my knowledge and was prepared under the AI governance framework approved by the Board.* Signed: ______________________ Role: ______________________ Date: __________

Template note: *This summary is designed to fit a single printed page at standard board report formatting. Values shown are illustrative. Replace with actuals from your AI asset register, TCAIO model, incident log, and accountability chain documentation.*

The Governance Record

The Board AI Governance Summary is not only a communication tool. It is a governance record. Every completed summary should be retained in the board's formal governance documentation, with the Executive Sponsor's signature and the date of board committee receipt. That retention record is the evidence of active board oversight that the Business Judgment Rule requires and that regulatory examiners, in an increasing number of industries, are beginning to request as a standard examination document.

THE DOCUMENT THAT CHANGED THE EXAMINATION

Illustrative Scenario: *The following scenario is representative of documented patterns in executive AI governance. Names, organizations, and specific details are illustrative.*

The regional bank had been preparing for its safety and soundness examination for three months. The technology team had assembled a documentation package that ran to several hundred pages: model validation reports, system architecture diagrams, vendor contracts, compliance assessments, training data documentation, and performance logs for every AI system in the organization's portfolio. It was, by any technical standard, a comprehensive package.

On the first day of the examination, the lead examiner asked the CEO a single question: "Can you show me the document your board uses to govern your AI program?"

The CEO called the CRO. The CRO pulled the board packet from the prior quarter. The AI section of the board packet was the twenty-three-page technology briefing. The lead examiner spent eight minutes reviewing it. She said: "This tells me what your systems do. It doesn't tell me how your board governs them. Do you have anything that shows me board-level accountability, risk posture, and incident oversight?"

The CRO said she would need to compile that information.

The examination team spent the better part of two days working through materials that had never been assembled in governance format: board minutes that mentioned AI in passing but did not document specific oversight actions, risk committee presentations that reported performance metrics without documenting governance decisions, incident records that had been handled operationally without board notification. The picture that emerged was of a technology function doing serious governance work and a board that had approved the existence of that work without exercising meaningful oversight of it.

The examination produced a Matters Requiring Attention finding: the bank's board did not have adequate visibility into the AI governance program's risk posture, incident history, or accountability structure. The finding was not about the AI systems themselves, none of which had malfunctioned. It was about the governance layer. The board was not seeing what it needed to see to exercise oversight.

The CRO spent the following quarter building the Board AI Governance Summary. She used the six-section framework, populated it with data already held by the technology, risk, legal, and finance functions, and had the first completed summary reviewed by the board risk committee sixty days after the examination findings were issued. The board member from this chapter's opening was on the risk committee. She spent fourteen minutes asking questions. She moved to approve the governance summary and requested it at every subsequent meeting.

At the next examination, eighteen months later, the lead examiner asked the same opening question. The CEO turned to the CRO. The CRO handed over a folder containing six consecutive Board AI Governance Summaries, each signed by the Executive Sponsor, each with the date of board committee receipt stamped on the cover page, each showing the governance program's evolution over six quarters: more systems added, one Tier 3 system reclassified from Tier 2, two incidents logged and closed, TCAIO trending toward budget.

The lead examiner reviewed the summaries for approximately twenty minutes. She asked four questions, all of which the summaries answered directly. She said: "This is the governance record we're looking for. Your board is informed. Your oversight is documented. We'll want to see this continue." The examination produced no findings related to AI governance.

The six governance summaries had taken approximately three hours each to produce. The examination finding from the prior cycle had taken more than two weeks of management time to respond to and had required an MRA remediation plan that consumed several months of governance resources. The math was not complicated. The governance summary was not a reporting exercise. It was risk management with a quarterly cadence.

PRODUCING THE FIRST SUMMARY THIS QUARTER

The first Board AI Governance Summary is the hardest one, because it requires assembling data from the asset register, the tier classifications, the TCAIO model, and the accountability chain documentation into a single format for the first time. Subsequent summaries are updates to a standing document. Four actions that produce the first one.

1. **Block four hours and assemble the source data before writing a single word.**

The six sections of the governance summary draw from four source documents built in prior chapters: the AI asset register (Sections 1, 2, and 5), the TCAIO model (Section 4), the accountability chain documentation (Section 5), and the incident log (Section 3). Section 6 requires a brief consultation with your General Counsel about the regulatory landscape. Before opening the governance summary template, assemble all four source documents and have your GC's regulatory assessment in writing. The summary is a synthesis document. The quality of the synthesis depends entirely on the completeness of the sources. If any source document is incomplete, note the gap explicitly in the corresponding summary section rather than leaving it blank or filling it with approximations.

Write from the source documents, not from memory. The Board AI Governance Summary is a synthesis document, and its value depends on the accuracy of the underlying data. Assemble the asset register, the tier classifications, the TCAIO summary, and the incident log before writing a single word of the summary. The writing takes two hours. Assembling the source data takes four. Allocate accordingly. Success looks like a governance summary where every figure, every system name, and every status designation is traceable to a source document.

2. **Write Section 3 first, then Section 2.**

The instinct is to complete the summary sequentially. Section 1 through Section 6. The governance logic runs in the opposite order. Incidents (Section 3) are the most time-sensitive information: if something went wrong this quarter, the board needs to know before they read the asset summary. Writing Section 3 first forces the author to think clearly about what happened, how it was handled, and whether anything remains open. Section 2 (Risk Posture) is the section that contextualizes Section 3, the incident log is more meaningful when the board can see it against the tier distribution and override rate data. Writing Section 2 second produces a risk posture assessment that reflects the quarter's actual experience rather than a generic status update.

Most common mistake: understating incidents in Section 3 to avoid difficult board conversations. The governance summary is a fiduciary record, and a governance record that minimizes or omits incidents is not merely incomplete, it is a misrepresentation that exposes the Executive Sponsor to the personal liability consequences the Business Judgment Rule protection is designed to prevent. Incidents that were detected, handled, and resolved are not governance failures. They are governance successes. Report them fully. Success looks like a Section 3 that a board member who had read none of the underlying documentation could understand completely.

3. **Have a non-technical board member read the draft before it goes to committee.**

Before the governance summary is distributed to the board committee, give it to a single board member who is not on the technology committee and does not have a technical background. Ask one question: can you tell me, based only on this document, whether there is anything about our AI program that you would want to ask management about in a board meeting? If the answer is no, meaning they found nothing to ask about, the document may be too reassuring in its language. If the answer is yes, but the question they want to ask is not one the document addresses, the document is missing something. The goal is a document that a non-technical board member can interrogate confidently. The only way to verify that the document achieves that goal is to give it to a non-technical board member before the meeting.

Rotate the pre-reader every quarter. The pre-reader's job is to tell you what a non-technical board member cannot understand and what questions the summary will generate. After three consecutive quarters with the same pre-reader, they stop reading as a board member and start reading as a governance insider. The rotation maintains the layperson perspective that makes the feedback useful. Success looks like a pre-reader who asks at least one question the Executive Sponsor cannot immediately answer from memory, because that is the question the committee will ask.

4. **Present the governance summary separately from the technology briefing.**

Many organizations that begin producing Board AI Governance Summaries embed them in the technology function's board presentation as an introductory slide or attach them as a cover page to the full technical report. This placement signals to the board that the governance summary is a preface to the real content rather than the governance document itself. Present the governance summary as a standalone agenda item at the risk committee, separate from and prior to any technology briefing. The governance summary gets its own agenda slot, its own discussion period, and its own approval motion. The technical briefing, if presented at all, is supplementary material. That sequence inverts the signal: governance first, technology detail second. The board is governing an AI program, not reviewing a technology report.

Separate the governance presentation from the technology briefing on the agenda, and present them in that order. The governance summary should stand alone, presented by the Executive Sponsor, before the technology team's capability update. This sequencing establishes governance as the frame within which capability is evaluated, not as an appendix to the technology narrative. Success looks like a board agenda that lists "AI Governance Summary" and "Technology Update" as separate items with separate presenters and separate discussion time.

THE DOCUMENT THAT MAKES OVERSIGHT VISIBLE

The board member from this chapter's opening had six years of governance experience and a genuine commitment to her fiduciary responsibilities. What she lacked was not knowledge or effort. It was a document that translated the organization's AI governance work into the terms that fiduciary oversight requires. The twenty-three-page technology briefing was not a failure of effort. It was a failure of format: information prepared for a technical audience, delivered to a governance audience, producing the appearance of oversight rather than the substance of it.

Board governance of AI is not about technical literacy, it is about decision-ready information, clearly structured, that enables a non-technical board member to ask good questions, make good decisions, and create a governance record that demonstrates active oversight. The Board AI Governance Summary is the instrument that makes that possible.

Without it, governance is happening somewhere in the organization. With it, governance is visible to the board, and the board's oversight of it is documented.

Part Two is complete. You have built the five instruments of the governance framework: the AI asset register, the risk tier model, the total cost of AI ownership model, the accountability chain, and the board governance summary. Each instrument changes how the organization sees its AI exposure. Together, they form a governance program that a board can interrogate, a regulator can examine, and an organization can sustain. In Part Three, the work shifts from building the framework to deploying it. Chapter Nine gives you the ninety-day implementation plan that takes you from the frameworks in this book to a governance program your board can see, your regulators can examine, and your organization can sustain.

> ***The board that cannot interrogate AI governance is not exercising AI governance. The governance summary is what makes interrogation possible, and oversight real.***

BOARDROOM TAKEAWAY | *The 90-Day Audit*

What a board member should be able to ask:

- Is there a governance calendar for the 90-day sprint with protected time blocks confirmed?
- By Day 90, will we have a completed register, a tiered risk assessment, a chain of accountability, and a board governance summary?

What a CFO or risk leader should have documented:

- A kickoff meeting scheduled and a four-person task force formed with named members
- The ten kickoff questions answered in writing, establishing the pre-sprint governance baseline

What evidence should exist:

- A Day 90 governance record: completed register, TCAIO summary, chain documentation, adopted policy, and board summary
- Board committee documentation confirming receipt of the first governance summary and the quarterly cadence established

CHAPTER NINE: PART III: THE EXECUTION

The 90-Day Audit

Going from Exposed to Accountable on a Real Calendar

She had read the book. She had attended the conference. She had commissioned two internal assessments and reviewed three consulting proposals ranging from $180,000 to $640,000. She had everything except a start date.

> **Illustrative Scenario:** *The following scenario is representative of documented patterns in executive AI governance. Names, organizations, and specific details are illustrative.*

The COO had been trying to launch her organization's AI governance program for fourteen months. Every time the launch approached, something intervened: a regulatory filing, a board retreat, a quarter-end close, a vendor contract renegotiation that consumed the legal team's capacity. Each delay was individually reasonable. Collectively, they had produced fourteen months of intention without a governance structure, fourteen months of AI deployment without a framework, and fourteen months of board meetings in which AI was discussed but not governed.

In the fifteenth month, her organization received a regulatory inquiry. It was not an enforcement action. It was a standard information request from a state regulator asking for documentation of the organization's AI governance practices. The COO had sixty days to respond.

She called me on the day the inquiry arrived. She said: "I need to build in sixty days what I have been trying to start for fourteen months. Tell me exactly what to do and in what order."

I told her the good news first: sixty days was enough to build a governance foundation that could respond to the inquiry honestly, completely, and in a way that demonstrated genuine organizational commitment rather than reactive assembly. The bad news was that she could not do it in sixty days by trying to do everything simultaneously. She needed a sequence, a team, a set of weekly commitments, and the discipline to protect the calendar against the next fourteen months' worth of reasonable interruptions.

We structured her sprint in sixty days because that was what the regulatory timeline required. This chapter gives you ninety, not because the work requires ninety days, but because ninety days allows you to do it right rather than fast, to build governance you can sustain rather than documentation you have to apologize for, and to arrive at Day 90 with a board governance record, not just an internal project completion.

The sprint begins the day you set the kickoff meeting.

THE GOVERNANCE PROGRAM THAT NEVER STARTS

The most common AI governance failure I encounter is not the organization that tried and failed to govern AI. It is the organization that is genuinely committed to governing AI and has been preparing to start for longer than it would take to finish. Every element of this description will be familiar to any executive who has been in the position: the internal assessments have been commissioned and reviewed.

The consulting proposals have been evaluated. The frameworks have been benchmarked. The leadership team has reached alignment on the importance of the work. The governance program itself has not been built.

This failure mode is not produced by negligence or avoidance. It is produced by perfectionism and sequencing. The governance program has not started because leadership is waiting for the right moment: the quarter without a competing priority, the team with full capacity, the regulatory environment with complete clarity, the internal alignment that is still two conversations away. That moment does not arrive. The competing priorities are structural, not cyclical. The team capacity is always partially consumed. Regulatory clarity is always partial. The waiting is not a temporary condition. It is the default condition of every organization trying to build a new governance function while running existing operations.

> ***The governance program that is almost ready to start is governed by the same gap as the governance program that has not started: the gap between intention and structure. This chapter closes it.***

The cost of the delay is compounding rather than static. Every month without a governance structure is a month in which AI systems are deployed without tier classification, accountability chain assignment, or TCAIO analysis. The inventory grows. The shadow AI accumulates. The vendor contracts that do not include retraining notification requirements get renewed without amendment. The board receives briefings rather than governance summaries. And when the regulatory inquiry arrives, or the litigation discovery request, or the board chair's question that cannot be answered, the organization is building the governance structure in the conditions least conducive to building it well.

The 90-Day Accountability Sprint does not require perfect conditions. It requires a kickoff meeting, a cross-functional task force, ninety calendar days, and the executive commitment to protect that calendar against the reasonable interruptions that will otherwise consume it. The sprint is not designed to be completed in the spaces between competing priorities. It is designed to be treated as a competing priority, one with a board-level governance consequence that outranks most of what will try to displace it.

THE 90-DAY SPRINT: THREE MONTHS, ONE GOVERNANCE FOUNDATION

The 90-Day Accountability Sprint is organized as three consecutive monthly sprints, each with four weekly milestones, a governance checkpoint at the end of the month, and specific task force deliverables that build on the prior month's work. Before the sprint begins, a kickoff meeting establishes the starting position and surfaces the gaps that define the Month One workplan. Before the sprint ends, two decisions require executive authority that the task force cannot provide. By Day 90, one document must exist that has not existed before.

The sprint is sequenced deliberately. Month One produces the inventory before anything is classified, so the classification work in Month Two is applied to a complete picture rather than a partial one. Month Two produces the risk tiers, cost models, and accountability chains before the board summary is drafted, so the summary in Month Three reflects actual governance data rather than estimated governance posture. Month Three produces the board summary before the sprint closes, so the governance work produces a record, not merely a project completion. The sequence matters. Compressing or reordering it produces a governance program that is formally complete and structurally unsound.

THE 10 KICKOFF QUESTIONS

Ask these questions at the kickoff meeting, before the sprint begins. The answers establish your starting position. The gaps they surface define the Month One workplan.

1. Can anyone in this room name every AI system currently running in our organization? If not, where does that list currently live, and who owns it?
2. Do we have a single document that classifies each system by its level of decision autonomy?
3. For any system that makes or heavily influences a decision affecting customers, employees, or financial outcomes, who is the named executive accountable for its governance?
4. What is our current total spend on AI, across all budget categories, including infrastructure, legal, compliance, insurance, and governance oversight?
5. Has our board ever received a document that shows our AI risk posture, incident history, and accountability structure in a single page?
6. If an AI system produced a harmful output today, who would be notified first, and within what timeframe would the board be informed?
7. Do our vendor contracts for AI systems require the vendor to notify us when the model is retrained or when a material performance issue is detected?
8. Have we assessed whether any AI system currently operating as advisory has drifted toward autonomous in actual practice?
9. What regulatory requirements are within 18 months of taking effect that apply to our AI portfolio, and where are we in preparing for them?
10. If a regulator asked for our AI governance documentation today, what would we hand them, and how long would it take to assemble it?

Governance benchmark: *An organization with strong pre-existing governance will answer 7 or more questions confidently. Most organizations answer 2 to 4. The gap between where you are and 10 is the 90-day sprint.*

MONTH 1, INVENTORY & CLASSIFICATION
By Day 30: A complete AI asset register. Every system named. Every tier assigned. Every shadow AI surfaced. No governance work begins until the inventory is complete.

Week 1. Task Force Formation & Detection Launch
Appoint the four-person cross-functional task force (IT, Legal, Procurement, Finance) with the Executive Sponsor confirmed. Launch all three detection passes simultaneously: spend analysis, vendor update log review, and technical staff survey. Set the governance calendar for the full 90-day sprint with protected time blocks. Brief the General Counsel on the sprint's scope and timeline.

Week 2. Detection Synthesis & Preliminary Register

Close all three detection passes. Merge results into a single master list. Begin populating the AI asset register five-column template. Identify every system whose accountable owner (Column 5) cannot be named, flag for Week 3 executive resolution. Deliver preliminary register draft to Executive Sponsor for review. Executive Decision 1 occurs this week: Tier 3 owner confirmations.

Week 3. Register Completion & Classification Workshop

Complete all five columns for every system in the register. Conduct one-hour classification workshops for each system whose Column 4 designation (decision authority) is unclear or contested. Apply the operational reality test: pull override rates for every Tier 1 and Tier 2 candidate. Document the two-dimension tier assessment for each system.

Week 4. Register Finalization & Month One Checkpoint

Finalize tier assignment for every system. Add the Accountability Chain column (seventh column) as a placeholder to be completed in Month Two. Executive Sponsor reviews completed register. Any system still missing a Column 5 owner is escalated to CEO authority before the Month One checkpoint closes.

GOVERNANCE CHECKPOINT

Executive Sponsor reviews the completed AI asset register. Key questions: Does this represent every AI system we believe is running? Are there detection gaps we should close before Month Two? Are all Tier 3 systems assigned to named executive owners? The month does not close until the register is complete and the Executive Sponsor has signed off on its accuracy.

EXECUTIVE DECISION REQUIRED

Tier 3 accountability owner confirmation. When the task force identifies a Tier 3 system with a disputed or missing Executive Sponsor, the CEO or equivalent must make the assignment explicitly, in writing, in the governance record. This decision cannot be delegated to the task force.

MONTH 2, RISK TIERING, COST MODELING & CHAIN ASSIGNMENT

By Day 60: Every system risk-tiered with documented rationale. TCAIO completed for all Tier 3 and highest-cost Tier 2 systems. Accountability chains fully populated for all Tier 2 and Tier 3 systems.

Week 5. Tier Model Application & Override Rate Verification

Apply the AI Risk Tier Matrix to every system in the register. Verify override rates against system logs (not self-report) for every Tier 1 and Tier 2 system in a consequential domain. Reclassify any system whose operational reality diverges from its designed designation. Document the two-dimension rationale for every tier assignment.

Week 6, TCAIO Completion for Tier 3 Systems
Convene a one-hour cost attribution meeting with Finance, Legal, and IT for each Tier 3 system. Populate all seven TCAIO categories for each Tier 3 system. Calculate the actual ROI ratio for each Tier 3 system against its original business case. Flag any system with an ROI ratio below 0.5 for Executive Sponsor review.
Week 7, TCAIO Completion for Tier 2 Systems & Chain Assignment
Complete TCAIO analysis for highest-cost and highest-consequence Tier 2 systems. Populate the accountability chain (all six links for Tier 3; Links 1 through 5 for Tier 2; Links 1 through 3 for Tier 1) in the register. Conduct Link 1 vendor contract audit for all Tier 2 and Tier 3 systems: identify the three contract provisions that are absent and require amendment.
Week 8. Chain Confirmation Workshops & Month Two Checkpoint
Conduct the two-hour chain confirmation workshop for each Tier 3 system. Confirm acceptance of governance role obligations by each named Link holder. Schedule tabletop exercise for at least one Tier 3 system before Day 90. Deliver Month Two deliverables to Executive Sponsor: completed tier register, TCAIO portfolio summary, and accountability chain documentation.

GOVERNANCE CHECKPOINT
Executive Sponsor reviews the complete tier classification, TCAIO portfolio summary, and accountability chain documentation. Key questions: Does the tier distribution reflect an honest assessment of how these systems actually operate? Are there systems with TCAIO results that require a board decision about their strategic justification? Is every Tier 3 system's accountability chain fully populated and confirmed by each Link holder? Any Tier 3 TCAIO result with an ROI ratio below 0.5 should be presented to the full executive team before Month Three begins.

MONTH 3, BOARD REPORT, POLICY ADOPTION & FIRST QUARTERLY REVIEW
By Day 90: The Board AI Governance Summary drafted, reviewed by a non-technical board member, and presented to the board committee. The AI governance policy formally adopted. The quarterly governance cadence established as a standing commitment.

Week 9. Governance Policy Drafting
Draft the organization's AI governance policy. The policy is a single document, typically four to eight pages, that defines the governance framework the sprint has built: the tier model, the TCAIO requirement for new deployments, the accountability chain structure, the board reporting cadence, the incident escalation protocol, and the vendor contract standards. The policy is not a technical document, it is an organizational commitment document. Draft it in the language of governance, not technology.

Week 10. Board Governance Summary Draft & Policy Review
Draft the first Board AI Governance Summary using the Chapter Eight template, populated with the actual data from the sprint. Executive Decision 2 occurs this week: governance policy adoption by formal executive or board resolution. Have the draft governance summary reviewed by one non-technical board member before Week 11.
Week 11. Tabletop Exercise & Summary Finalization
Conduct the tabletop exercise for the highest-consequence Tier 3 system. Document the exercise, its participants, and its findings in the governance record. Incorporate the tabletop findings into the governance summary's Section 5 (Accountability Status). Finalize the Board AI Governance Summary based on board member pre-review feedback.
Week 12. Board Presentation, Policy Adoption & Cadence Commitment
Present the Board AI Governance Summary to the risk or audit committee as a standalone agenda item, with the Executive Sponsor as presenter. Confirm formal adoption of the AI governance policy. Establish the Board AI Governance Summary as a standing agenda item at every quarterly board committee meeting. Document the Day 90 governance record: completed register, TCAIO summary, chain documentation, adopted policy, first board summary, board receipt confirmation.

GOVERNANCE CHECKPOINT
The board committee receives the first Board AI Governance Summary and the adopted governance policy. Key questions: Does the board have the information it needs to exercise meaningful oversight of the AI portfolio going forward? Are there any findings from the 90-day sprint that require a board decision before the next quarterly meeting? Is the governance program's quarterly cadence confirmed on the board calendar? The sprint does not close until the board committee has received and documented its receipt of the governance summary.
EXECUTIVE DECISION REQUIRED
Governance policy adoption by formal executive or board resolution. A governance policy that exists in a risk management folder without formal adoption does not create organizational accountability. The adoption record, board resolution or executive order, is the document that converts the policy from a recommendation into an organizational commitment.

The Two Decisions That Require Executive Authority
Decision 1 (Week 2): *Tier 3 accountability owners must be confirmed by executive authority, not proposed by the task force. When the task force identifies a Tier 3 system with no named Executive Sponsor, or when there is organizational disagreement about which C-suite role carries the governance accountability, that decision cannot be resolved by the people doing the inventory work. It requires the CEO, CFO, or equivalent to make an explicit, documented assignment. Without*

executive authority behind the assignment, the accountability is nominal. It will not hold under incident pressure.

Decision 2 (Week 10): *The governance policy must be adopted by formal executive or board resolution, not implemented by the technology or risk function alone. A governance policy that lives in a risk management folder and has never been formally adopted does not create organizational accountability, it creates documentation. The policy must be adopted at the level that gives it organizational authority: board resolution for policies with board-level governance implications, executive order for operational governance policies. The adoption document is the record that the organization made a deliberate governance decision, not just a project completion.*

THE ONE DOCUMENT THAT MUST EXIST BEFORE DAY 90

The Board AI Governance Summary from Chapter Eight, first edition.
Not the perfect version. Not the fully-populated, every-gap-closed version. The first version: accurate, honest about what is still in progress, signed by the Executive Sponsor, and presented to the board committee. That document is the proof of concept that the 90-day sprint was a governance act, not a project. It converts the prior 89 days of work into an oversight record. It gives the board something to interrogate, approve, and request at the next meeting. It establishes the cadence that makes AI governance sustainable beyond Day 90. A sprint that produces every governance instrument except the board summary has produced everything except the evidence that governance happened. The board summary is the governance record that makes the rest of the work visible and the accountability real.

SIXTY DAYS ON A REGULATORY CALENDAR

The COO whose regulatory inquiry opened this chapter completed her governance sprint in fifty-eight days. The conditions were not ideal, they were real, and the outcome was achieved despite them, not because of them.

On Day One, she held the kickoff meeting with six people in the room: the CTO, the Chief Risk Officer, the General Counsel, the CFO, the head of IT, and the VP of Operations who owned the organization's two largest AI-dependent business processes. She asked nine of the ten kickoff questions. The tenth, about the regulatory documentation they could produce today, was the question that had prompted the meeting, and everyone already knew the answer.

The answers to the other nine questions established the starting position precisely. The organization could name approximately half of its AI systems with confidence. It had no tier classification framework. It had no TCAIO model. It had accountability for AI systems distributed across the CTO's function without named executive-level owners for the two systems whose consequence magnitude was clearly Tier 3. The board had never received a governance summary. The vendor contracts had been reviewed for commercial terms but not for governance provisions.

She framed the starting position in a single sentence before the meeting ended: "We have AI running and governance that hasn't caught up with it. In fifty-eight days we are going to close that gap, produce a documentation package for the regulator that reflects genuine governance, and present a board summary that we can stand behind. This is not a documentation project. It is a governance project that produces documentation. The difference matters."

Month One ran on schedule because the task force was small, the detection passes were run in parallel, and the Executive Sponsor made herself available for exactly one thirty-minute decision meeting in Week Two: the Tier 3 accountability owner assignments. There were three Tier 3 systems. Two had natural organizational homes, the CFO for the financial risk model, the Chief Operating Officer for the autonomous logistics routing tool. The third was a customer service AI that had been built internally and whose outputs affected customer-facing decisions across three business lines. The COO took that one herself. The assignment was documented in the meeting minutes before the meeting ended.

Month Two surfaced the sprint's most significant finding. The TCAIO analysis for the autonomous logistics routing tool, which had a subscription cost of $180,000 annually and had been presented to the board as a high-ROI investment, produced a total annual cost of $620,000 when infrastructure, compliance, legal review, oversight staffing, and the liability reserve were included. The reported ROI ratio was 2.1 times. The actual ROI ratio was 0.6 times. The tool was cash-flow negative.

The COO presented this finding to the executive team in the Week Eight checkpoint meeting. The reaction was what it always is when a TCAIO analysis produces a number significantly larger than the subscription cost: a period of attempted disproof, followed by acceptance, followed by a decision about what to do. The executive team decided to retain the tool, it had operational strategic value beyond its direct financial performance, but to reclassify it as a strategic investment rather than a cost-efficiency tool in all future board presentations. The governance record for that decision was documented. The board would see it in the governance summary.

The Day 58 board committee meeting ran thirty-two minutes. The COO presented the governance summary. The committee chair asked five questions, all of which the summary answered directly. The committee approved the governance policy by formal resolution. The committee chair asked that the summary be provided at every subsequent quarterly meeting and noted in the minutes that the organization had "demonstrated a genuine commitment to building AI governance infrastructure, not merely producing documentation responsive to a regulatory inquiry."

The regulatory response was filed on Day 60. It included the completed AI asset register, the tier classification documentation, the TCAIO portfolio summary, the accountability chain records for all Tier 3 systems, the adopted governance policy, and the board committee meeting minutes that documented receipt and approval of the first Board AI Governance Summary. The regulator acknowledged receipt. Four weeks later, the inquiry was closed with no findings.

The COO told me afterward that the experience had changed how she thought about governance preparation. Not because the regulatory outcome was favorable, though it was. Because the governance program the sprint had built was not a regulatory response. It was a governance structure that the organization now had, would continue to maintain, and could demonstrate at any future examination, board meeting, or leadership transition. The regulatory inquiry had been the forcing function. The sprint had been the method. What remained was a governance foundation.

STARTING THE SPRINT THIS WEEK

The sprint begins with a calendar decision, not a governance decision. The governance decisions follow. Four actions that launch it.

1. **Schedule the kickoff meeting before you finish reading this chapter.**

The most important action in this chapter is not a governance action. It is a calendar action. The kickoff meeting needs six people: the Executive Sponsor, the CTO or equivalent technology leader, the Chief Risk Officer, the General Counsel, the head of IT, and the operational leader most accountable for the highest-stakes AI deployment in the organization. The meeting is ninety minutes. It ends with a confirmed task force, a confirmed Day 90 date, and the ten kickoff questions documented with their answers. Schedule it before the competing priorities of next week make it the fifteenth month.

Schedule the kickoff and protect the ninety days simultaneously. The kickoff meeting is the governance commitment made visible. The protected calendar blocks are the governance commitment made real. An organization that holds the kickoff meeting and then allows the sprint work to compete with normal priority demands will arrive at Day 90 with a partially completed governance program and a board disclosure problem. Block the time before the kickoff, not after it. Success looks like a shared calendar where the sprint's weekly working sessions are immovable for three months.

2. **Appoint the Executive Sponsor and clarify what that role requires.**

The Executive Sponsor is not the person who commissioned the sprint, approved the budget for it, or expressed interest in AI governance at the last leadership offsite. The Executive Sponsor is the C-suite leader who will sign the Board AI Governance Summary at Day 90, whose name will appear in the AI asset register as the accountable owner for Tier 3 systems, and who will present the governance summary to the board committee as a fiduciary document. That person needs to know, before the kickoff meeting, that this is what the role requires. The sprint produces governance accountability, and governance accountability requires a named person whose professional obligation includes owning it.

Most common mistake: naming the CTO or the head of risk management as the Executive Sponsor because they are the people closest to the work. The Executive Sponsor must be a C-suite leader with fiduciary authority, a CFO, COO, CEO, or equivalent. The governance summary the Executive Sponsor signs is a board governance record. It carries the weight of the fiduciary obligation that Chapter Three established. A director or VP cannot sign a board governance record in a way that activates the Business Judgment Rule's protection. The Executive Sponsor must be the right person, not the most available person. Success looks like a named C-suite Executive Sponsor who has explicitly accepted the role and its accountability obligations.

3. **Identify the one governance gap that, if closed in the next thirty days, would most reduce your organization's current AI exposure.**

Before the kickoff meeting, take thirty minutes to answer this question personally: based on what you know today about your organization's AI portfolio, what is the single governance gap whose closure would most reduce exposure to the four liability categories from Chapter One? It might be the unassigned Tier 3 system. It might be the vendor contract with no retraining notification clause. It might be the board that has never received a governance summary. It might be the shadow AI that has never been inventoried. Name it. Write it down. Bring it to the kickoff meeting as your starting hypothesis for Month One's highest-priority work. The task force will surface more gaps, but entering the sprint with a personal risk hypothesis focuses the Executive Sponsor's attention in the way that a generalized commitment to governance improvement does not.

Answer this question in writing before the kickoff meeting, not in the meeting itself. The Executive Sponsor who arrives at the kickoff without a written answer to "what is the one governance gap most urgent to close in thirty days" has already made the sprint's first governance error: a consequential decision made without prior analysis. Write the answer. It will shape everything that follows. Success looks like a kickoff meeting that opens with the Executive Sponsor naming the sprint's highest priority from a position of prior deliberation, not live reasoning.

4. **Tell your board chair that the sprint has begun.**

Before Day 7, notify your board chair or the chair of the relevant board committee that the organization has launched a ninety-day AI governance sprint and that the first Board AI Governance Summary will be presented at the committee meeting closest to Day 90. This notification serves three governance purposes. First, it creates a board-level accountability commitment that the Executive Sponsor cannot quietly defer without notifying the board chair. Second, it signals to the board that AI governance is being treated as a fiduciary priority rather than a technology project, which is exactly the reframe that Chapter Three established as the appropriate governance posture. Third, it gives the board chair the opportunity to ask questions, signal priorities, and engage with the governance work before the Day 90 presentation, which produces a better governance summary because the Executive Sponsor knows what the board is most focused on.

Brief the board chair before Day 90, not on Day 90. A board that receives its first governance summary with no prior context experiences it as a disclosure. A board whose chair has been briefed informally at Day 30 and Day 60 experiences it as a completion. The difference is not in the document. It is in the relationship it represents. Success looks like a board chair who arrives at the Day 90 presentation already understanding why the sprint was necessary, what it produced, and what the quarterly governance cadence will require of the committee going forward.

THE FOUNDATION, NOT THE FINISH LINE

The COO whose sprint opened this chapter told me, two years after her sixty-day sprint, that the sprint had been the best governance investment her organization had made in her tenure. Not because of the regulatory outcome, though that had been favorable. Not because the governance program had prevented an incident, though it had surfaced and resolved two in the eighteen months since. Because the sprint had converted her organization from one that was intending to govern AI into one that was actually governing it, and the difference between those two states was visible to her board, verifiable by her regulator, and felt by every member of her executive team who had participated in building it.

The 90-day audit is not a transformation, it is the foundation of one. By Day 90, you will have a governance structure that did not exist before, a board record that demonstrates you built it deliberately, and a quarterly cadence that makes the governance sustainable past the sprint that launched it. The structure is the beginning. What you do with it in the quarters that follow is the actual governance program.

The 90-day sprint is not the governance program. It is the proof that the governance program is real. By Day 90, your organization will have something it did not have before: a board record showing that governance was built deliberately, a register that names every AI

system, a chain that names every accountable owner, and a quarterly cadence that keeps all of it current. That proof is what Chapter Ten prepares you to deliver.

> ***By Day 90, you will have a governance structure that did not exist before, and a record that demonstrates you built it deliberately, rigorously, and with board-level transparency.***

BOARDROOM TAKEAWAY | *The Board Room Moment*

What a board member should be able to ask:

- Can you deliver all five governance statements from memory, without notes, in under ten minutes?
- Does your board currently have a governance summary they can interrogate, or only a capability briefing?

What a CFO or risk leader should have documented:

- The five prepared governance statements, practiced and rehearsed before the board meeting
- Supporting documentation for each statement: the register, the chain, the TCAIO model, the incident protocol, and the readiness matrix

What evidence should exist:

- Board meeting minutes documenting that governance statements were made and received
- A record distinguishing between governance and capability content in all AI-related board communications

CHAPTER TEN: PART III: THE EXECUTION

The Board Room Moment

Read this sentence aloud before continuing: 'Our AI inventory currently tracks [number] systems. [Number] are Tier Three, meaning they make or heavily influence consequential decisions without direct human approval at each step. Our highest-risk system is [system name], owned by [executive name]. Its current override rate is [rate], which is [within / outside] our governance threshold.'

If you can say that sentence, and four more like it, without notes, from the actual governance data your organization maintains, you are the executive this chapter confirms. If you cannot, this chapter is about why the knowledge is not the problem. The delivery is.

Saying It With Confidence, Specificity, and Authority

The CFO knew the answer. He was not sure he could say it.

He had done the work. The AI asset register was complete. The tier classifications had been reviewed and confirmed. The TCAIO analysis had been finished two weeks before the board meeting. The accountability chain was documented for every Tier 2 and Tier 3 system. He had been the Executive Sponsor for nine months. He had signed the governance summary. He knew what was in it.

And yet, seated at the board table with the risk committee chair looking at him across a glass of water and asking "Can you give us a picture of where we stand on AI governance?", the CFO felt the particular vertigo of a person who has all the information and none of the words.

He reached for his notes. He opened his laptop. He said: "I'll pull up the summary." The committee chair said, gently: "I've read the summary. I'm asking you."

What followed was four minutes of technically accurate, organizationally complete, and governance-credible information delivered in a sequence that produced no discernible impression of command. He covered the register, the tier distribution, the cost model, the incident log, and the regulatory horizon. Every number was correct. Every finding was disclosed. When he finished, the committee chair said: "Thank you. That's helpful." And moved to the next agenda item.

"That's helpful" is not what a board member says when she has received governance information that she trusts and intends to act on. It is what a board member says when the information has not been made interrogable.

The CFO called me the next morning. He said: "I had everything I needed. I couldn't make it land. What did I miss?" This chapter is the answer.

THE GAP BETWEEN KNOWING AND GOVERNING

There is a specific and recurring failure mode in board presentations on AI governance, and it does not belong to executives who are unprepared. It belongs to executives who are thoroughly prepared and have not organized their preparation into a form that produces board confidence. The failure is not informational. It is architectural. The executive knows what they know, but they have not built the structure that converts knowing into governing in real time.

Board governance operates through statements, not data. When a board member asks about a company's financial position, the CFO does not recite the balance sheet line by line. They make a statement: "We are well-capitalized, with $180 million in available liquidity against a debt maturity schedule that creates no material pressure in the next thirty-six months." That statement is supported by the balance sheet. It is not the balance sheet. The statement is what governance sounds like. The balance sheet is what governance is built on.

AI governance has the balance sheet, the register, the tier model, the TCAIO, the chain, the summary. What most executives do not yet have are the statements. The five governance statements in this chapter are the AI governance equivalent of the CFO's liquidity summary: structured, specific, evidence-backed, and designed to be delivered without notes in a room where someone is watching to see whether you actually know what you are governing.

> ***The board does not need your data. It needs your governance. The five statements are the form that governance takes when it is ready to be interrogated.***

Not having the statements is a governance credibility gap that accumulates over board meetings. A board that has seen its CFO open a laptop to answer an AI governance question in three consecutive meetings has registered, consciously or not, that the Executive Sponsor does not have command-level knowledge of the governance program. That registration affects trust, affects the quality of board engagement with AI governance, and affects the board's willingness to approve AI investments and governance postures that the executive team is recommending. The statements are not a communication upgrade. They are a governance instrument.

THE FIVE STATEMENTS: GOVERNANCE AT THE SPEED OF CONVERSATION

The Board AI Governance Summary is the evidence base. The five statements are what the executive says before handing the summary to the room. Each statement corresponds to a section of the summary, every number it cites lives somewhere in the six-section document from Chapter Eight. That alignment is deliberate: a board member who hears the same structure at every quarterly meeting learns to listen for the parts that have changed. That is what governance attention looks like from the board side of the table.

Each statement is written in a template script, not because the words must be exact, but because the structure must be consistent. Practice until the structure is automatic. Then the specific numbers are the only thing that changes quarter to quarter.

STATEMENT 1, OUR AI INVENTORY

WHY THIS STATEMENT MATTERS

This statement establishes that the executive has a complete and current picture of what AI is running in the organization. It signals that the governance program begins with known, classified, owned systems, not assumption and approximation. For most boards, this is still the most surprising statement in the sequence, because most boards have never received it.

SCRIPT

> ***"We are currently running [number] AI systems across the organization. [Number] are advisory, they inform human decisions. [Number] are assistive, they drive human decisions with a human confirmation step. [Number] are autonomous, they make or execute decisions without per-transaction human review. Every system has a named accountable owner. Since our last report, we added [number] and decommissioned [number]. There are no unclassified systems and no Tier 3 systems without an executive-level owner."***

SUPPORTING EVIDENCE

Completed AI asset register (Chapter Four). Current tier distribution from the governance summary Section 1. Any additions or decommissions since last report documented.

THE STATEMENT THAT SIGNALS UNPREPAREDNESS

"We have a number of AI tools in use across the organization, though we're still working to get a complete picture."

STATEMENT 2, OUR RISK POSTURE

WHY THIS STATEMENT MATTERS

This statement moves from inventory to exposure. A board member who knows what you have wants to know what keeps you up at night about it. The risk posture statement names your highest-exposure systems specifically, describes the oversight they receive, and gives the board a current direction of travel: is risk rising, stable, or declining? That directional signal is what board members use to decide how much attention the topic deserves.

SCRIPT

> ***"Our risk posture is [stable / improving / elevated] this quarter. Our [number] Tier 3 systems are [name them]. Each receives quarterly output reviews with statistical sampling and board-level reporting. Since our last report, there have been [no / the following] tier reclassifications. Our highest current governance concern is [name the specific system or issue], here is what we are doing about it and what the board needs to know."***

SUPPORTING EVIDENCE
Current tier distribution and any reclassifications from governance summary Section 2. Override rate data for any system flagged in the current period. The specific system or issue that represents the highest current governance attention.
THE STATEMENT THAT SIGNALS UNPREPAREDNESS
"We feel good about our risk posture overall. The team is monitoring things closely."

STATEMENT 3, OUR COST POSITION

WHY THIS STATEMENT MATTERS
This statement demonstrates that the executive has applied the discipline of full-cost visibility to the AI portfolio, that they are not managing AI costs the way most organizations manage them, through the technology budget line alone, but through the complete TCAIO model that shows the board what AI is actually costing. The ROI ratio is the statement's most powerful element: it gives the board a single number that tells them whether the AI portfolio is performing as approved.
SCRIPT

> ***"Our total AI cost on a fully-loaded basis is [TCAIO total] annually, against a technology budget line of [subscription cost]. The difference represents infrastructure, governance, compliance, insurance, and oversight costs distributed across other organizational budgets. Our actual ROI ratio, actual returns against total cost, is [ratio], compared to the [higher ratio] projected in the business cases we approved. The programs performing above their projected ratio are [name them]. The programs that require a strategic justification decision are [name them, if any]."***

SUPPORTING EVIDENCE
TCAIO portfolio total from Chapter Six model. Technology budget subscription line for comparison. Actual ROI ratio for the portfolio and for any individual system below 0.5. Any system requiring a board decision about strategic justification.
THE STATEMENT THAT SIGNALS UNPREPAREDNESS
"Our AI spend is within budget. We're seeing strong returns across the program."

STATEMENT 4, OUR ACCOUNTABILITY STRUCTURE

WHY THIS STATEMENT MATTERS
This statement answers the question that every board member is carrying but rarely asks directly: if something goes wrong, who is responsible? The accountability statement names that person for every high-stakes system. It describes the chain that connects the vendor contract to the board table. And it confirms that the chain has been tested, not just designed, which is the governance detail that distinguishes organizations that have built accountability structures from organizations that have documented them.

SCRIPT

> ***"Every Tier 2 and Tier 3 system has a named accountability chain from the vendor contract through the board committee. For our Tier 3 systems, the Executive Sponsors are [name and role for each]. The chain has been tested through tabletop exercises for [system names]. Our vendor contracts for Tier 3 systems have been reviewed and amended where needed, the retraining notification provision is now in place for [number] of our [total] Tier 3 vendor contracts."***

SUPPORTING EVIDENCE

Accountability chain documentation from Chapter Seven. Tabletop exercise records. Vendor contract audit results from Month Two of the 90-day sprint.

THE STATEMENT THAT SIGNALS UNPREPAREDNESS

"We have strong processes in place and clear ownership across the team."

STATEMENT 5, OUR REGULATORY READINESS

WHY THIS STATEMENT MATTERS

This statement positions the organization as a governance actor rather than a regulatory reactor. The board member who hears this statement should come away understanding that the executive team has visibility into what is coming, has assessed the organization's current posture against it, and has a preparation plan with a timeline. What they should not hear is uncertainty, vagueness, or an implicit request for the board to wait and see.

SCRIPT

> ***"The regulatory developments most relevant to our AI portfolio are [name one to three specifically, with their status]. Our preparation status is [describe each: not started / in progress / complete]. The development most likely to require a board decision before our next meeting is [name it, or confirm that none currently meets that threshold]. We are monitoring [additional developments] at the horizon that we will brief the board on as they mature."***

SUPPORTING EVIDENCE

Regulatory horizon section from governance summary (Section 6). General Counsel's current regulatory assessment. Any regulatory examination, inquiry, or enforcement action involving AI systems.

THE STATEMENT THAT SIGNALS UNPREPAREDNESS

"The regulatory landscape is evolving quickly and we're keeping a close eye on developments."

The Standard for Each Statement

Deliver each statement in under ninety seconds. ***If you need more than ninety seconds, you are reporting rather than governing.***

The governance summary is the report. The statement is the executive's synthesis of the report into a form the board can immediately interrogate. Practice each

> *statement aloud, without notes, against a ninety-second timer. The first time, you will use the notes. The fifth time, you will not need them.*

THE TEN BOARD QUESTIONS, AND THE ANSWERS THAT DEMONSTRATE FIDUCIARY COMMAND

Practice each answer aloud. The standard is not memorization. It is fluency.

BOARD QUESTION	THE ANSWER THAT DEMONSTRATES FIDUCIARY COMMAND
What AI systems are we running that I should personally know about?	Name the Tier 3 systems by name and function. Confirm each has a named Executive Sponsor accountable for its governance. The board member is asking whether there is anything material that should be in their oversight awareness, answer at the level of governance significance, not exhaustive inventory.
How do we know if one of our AI systems is producing biased outputs?	Describe the output review cadence for Tier 3 systems: quarterly statistical sampling, demographic pattern analysis, override rate monitoring. Name the Risk Officer who owns that review process. Confirm that the last review occurred on schedule and produced no material findings, or disclose any findings transparently.
If an AI system failed today, what would happen in the first 24 hours?	Describe the accountability chain escalation protocol: Technical Owner detects and notifies Business Owner and Risk Officer within 24 hours; Risk Officer escalates to Executive Sponsor; Executive Sponsor notifies the board committee if the event meets material threshold. Name the system you would be most concerned about and confirm its chain is populated.
What is AI actually costing us, total, not just the subscriptions?	Cite the TCAIO portfolio total and its relationship to the technology budget line. Name the largest cost category. Give the actual ROI ratio for the portfolio versus the reported ROI from business cases. This question is a financial governance question, not a technology question, answer in financial terms.
Are we exposed to the EU AI Act, or any other regulation that's coming?	Describe the regulatory horizon from the governance summary: which regulations are active, which are approaching, and your current preparation status for each. Confirm that Legal is engaged on the most material upcoming requirement. Do not speculate about regulatory outcomes, describe the preparation posture and the timeline.

Who is accountable when an AI system makes a decision that harms someone?	Name the Executive Sponsor for the relevant system by role. Describe the accountability chain from vendor contract through board committee. Confirm that the vendor contract specifies incident cooperation obligations. The board member is asking whether anyone specific has accepted organizational accountability, give them a name and a role, not a framework description.
How often does our board receive AI governance reporting?	Confirm the quarterly cadence of the Board AI Governance Summary. Name the committee that receives it. Describe what the summary covers, the six sections from Chapter Eight. If the cadence is not yet established, be direct about where you are in building it and when the first summary will be delivered.
Has anything gone wrong with our AI this year?	Answer from the incident log in the governance summary, completely and accurately. Describe any incident that occurred, how it was detected, how it was resolved, and whether any third-party harm resulted. A board member who learns about an AI incident from a source other than the Executive Sponsor's governance report has lost confidence in the governance program. Disclose proactively and fully.
How do we compare to what other organizations in our industry are doing?	This question often masks a concern about whether the organization is ahead of, at pace with, or behind the governance standard. Answer in terms of the governance structures that regulators and institutional investors are beginning to expect: AI asset register, tier classification, TCAIO model, accountability chain, and board reporting cadence. Describe which of these your organization has and where gaps remain.
What decision does this committee need to make today?	Every governance summary presentation should end with a clear answer to this question, identified in advance of the meeting. If the summary reveals a system whose ROI ratio requires a strategic justification decision, name it. If a regulatory development requires board approval of a preparation plan, present it. If no decision is required, say so explicitly: 'No decision is required this quarter; the summary is for oversight and information.' A board that knows when it is and is not being asked to decide is a board that can govern with confidence.

THE SECOND PRESENTATION

> **Illustrative Scenario:** *The following scenario is representative of documented patterns in executive AI governance. Names, organizations, and specific details are illustrative.*

The CFO from this chapter's opening had his second AI governance presentation to the risk committee four months after the first. In the intervening period, he had done three things. He had practiced each of the five statements in a thirty-minute weekly session with his Chief Risk Officer, using actual data from the governance summary as the supporting evidence. He had reviewed the ten board questions and prepared specific, documented answers to each one. And he had asked the committee chair, informally, what questions she most wanted to be able to ask in the next meeting, an act of governance preparation that also happened to be an act of relationship management.

She had told him she wanted to understand two things: the specific identity of the organization's autonomous systems, and what happened in the accountability chain the last time something went wrong. He had both answers ready.

The second presentation lasted nineteen minutes. He delivered all five statements without notes. He named both Tier 3 systems by name, described their decision domains, and confirmed the Executive Sponsor for each. He described the one incident from the prior quarter, a vendor retraining notification that had been received and reviewed without requiring escalation beyond Link 3, and walked the committee through the accountability chain's response step by step. He cited the TCAIO portfolio figure and the actual ROI ratio. He described the one regulatory development that was approaching a decision point and asked the committee to approve the preparation plan that Legal had developed.

The committee asked seven questions. He answered six of them from memory. The seventh, a specific question about the demographic breakdown of the output review results for one Tier 3 system, he answered by referring to Section 2 of the governance summary and pointing to the relevant line. He said: "I don't carry that level of detail in my head, but it's in Section 2 and the finding was clean." That answer, which acknowledged a limit while demonstrating command of where the information lived, was more confidence-producing than a memorized demographic breakdown would have been.

The committee chair moved to approve the regulatory preparation plan. Before the vote, she said: "I want to note for the record that the quality of this governance presentation has improved materially since last quarter. I believe we are receiving the information we need to exercise meaningful oversight of this portfolio. I am prepared to approve."

The committee approved four to zero in six minutes. The CFO had the same information in both presentations. What he had in the second presentation that he had not had in the first was structure: the five statements delivered in sequence, the supporting evidence cited by section, the questions anticipated and answered before they arrived.

The information had not changed. The governance had.

PREPARING FOR THE BOARD ROOM MOMENT

The five statements become fluent through practice, not study. Three actions that build the fluency.

1. **Draft all five statements from your actual governance data and read them aloud today.**

Using the scripts in this chapter as templates and your completed governance summary as the data source, draft all five statements in your own voice. Do not use the script language verbatim, the structure is the template, but the words should sound like you. Then read them aloud, to yourself, in sequence. The exercise will take approximately fifteen minutes the first time. What you will notice: two or three of the statements will feel natural immediately, because you are already fluent in the underlying data. One or two will feel uncertain, because you will reach for a number or a name and not have it immediately available. Those are the governance gaps your practice sessions should address.

Draft the statements, then read them aloud before the meeting. The board room is an oral environment. A statement that reads with authority on a page may stumble when spoken, especially under pressure. The statement that must be delivered without notes, in response to an unexpected follow-up question, is tested by speaking it, not reading it. Rehearse standing up. Rehearse with interruptions. Success looks like five statements you can deliver from memory in four minutes, with room for three follow-up questions you did not anticipate.

2. **Anticipate the two questions most likely to come from your specific board.**

The ten questions in this chapter represent the most common lines of board inquiry on AI governance. Your board has specific concerns, specific priorities, and specific lines of questioning that are predictable from your industry, your regulatory environment, and your board composition. Before your next board meeting, identify the two questions your board is most likely to ask that are not directly answered by the five statements. Prepare a documented answer to each. The documentation is not a script, it is a reference: a note card or a single page in your board briefing folder that you can cite if the question goes deeper than your prepared answer can take it.

Most common mistake: preparing for the questions you are comfortable answering rather than the questions you are least prepared for. The board member most likely to ask a difficult question about AI governance is typically the one with the most relevant domain expertise: the former financial regulator, the operations executive, the institutional investor representative. Think about who that person is on your board and what they are most likely to ask. Prepare for that question specifically. Success looks like two documented answers to board-specific questions, reviewed by your Chief Risk Officer before the presentation.

3. **Run a dry run with someone who will push back.**

Before the board presentation, deliver all five statements to a single person who has authority to challenge them: your CFO if you are the COO, your General Counsel, your Chief Risk Officer, or a trusted colleague with board-level experience. Ask them to ask three of the ten questions after you deliver the statements. The purpose is not to rehearse answers, it is to practice the transition from prepared statement to responsive governance in real time. A board member will not wait for you to finish Statement Five before asking a question about Statement Two. The dry run simulates the interruption pattern of an engaged board and builds the fluency that prepared statements alone cannot produce.

The dry run is a stress test, not a rehearsal. If the person asking questions is being cooperative, the dry run is not working. The point is to surface the question you cannot answer from current governance data, the gap in the narrative, the number you do not know. Those gaps, found in the dry run, are closed before the committee meeting. Found in the committee meeting, they become the record. Success looks like a dry run that ends with a short list of governance data you need to confirm before presenting, because that list is precisely what the dry run was designed to produce.

THE EXECUTIVE THE BOARD TRUSTS

The CFO from this chapter's opening had the same governance program in both presentations. The governance summary was the same quality. The underlying data was equally complete. What changed between the first presentation and the second was not the governance, it was the executive's ability to make the governance visible, specific, and interrogable in real time. The committee chair's statement for the record, "I believe we are receiving the information we need to exercise meaningful oversight", was not a comment about the quality of the governance summary. It was a comment about the quality of the executive's command of it.

Board readiness on AI governance is not about knowing everything, it is about knowing the right things with enough specificity to govern in real time: to answer the question you were not expecting and demonstrate to a room of experienced professionals that you are the executive whose judgment about AI they can trust. The five statements are the structure that makes that possible. The ten questions are the test.

The board member who leaves a meeting satisfied that governance is real, visible, and led by someone who commands the subject, that board member was created by the five statements and the ten questions in this chapter. You now have both. The next test is not a board meeting. It is something harder: the moment a system fails on your watch and the governance program is called upon to function under real pressure. Chapter Eleven is about that moment.

> ***The executive who can make the five statements without notes is not the one who knows the most about AI. They are the one the board trusts to govern it.***

BOARDROOM TAKEAWAY | *When It Fails on Your Watch*

What a board member should be able to ask:

- If an AI system produced a harmful output tonight, who receives the first call, and what is the documented escalation sequence?
- Has our incident response protocol been tested in a tabletop exercise before an actual incident?

What a CFO or risk leader should have documented:

- A completed AI Incident Response Protocol covering all seven steps
- A tabletop exercise record for the highest-consequence Tier 3 system, conducted before any actual incident

What evidence should exist:

- A documented escalation tree with names, roles, and timeframes at each step
- Board notification procedures specified in writing, including the trigger threshold and maximum time to board communication

CHAPTER ELEVEN: PART III: THE EXECUTION

When It Fails on Your Watch

The Incident Response Protocol You Build Before You Need It

She found out from a customer complaint.

Not from a system alert. Not from the Technical Owner. Not from the vendor. From a customer who had called the main line to say that the automated benefits determination the AI had produced did not match what a human reviewer had told him six weeks earlier, and that three of his colleagues had received the same conflicting determinations and were now disputing the outcomes with their case managers.

The COO took the escalated call herself. She asked her assistant to find out who was accountable for the benefits determination system. Her assistant came back twelve minutes later with the name of a director in operations who was on medical leave. She asked who was covering. Her assistant did not know. She asked the Technology team when the system had last been reviewed. They said quarterly reviews were scheduled but the most recent one had been deferred because of a product launch. She asked the vendor whether there had been any model changes. The vendor said they would need to check.

> **Illustrative Scenario:** *The following scenario is representative of documented patterns in executive AI governance. Names, organizations, and specific details are illustrative.*

In the four hours between the customer complaint and the end of the business day, seven more complaints arrived through three different channels. The COO had still not determined whether she was dealing with a model failure, a data error, a process gap, or something in the vendor's infrastructure. She did not know how many customers had been affected. She did not know whether there was a regulatory notification obligation. She did not know who was supposed to be sitting in the chair she was now occupying.

She called me that evening. She said: "I need a protocol. Not for next time. For right now. Walk me through what I should have had."

This chapter is what she should have had.

THE INCIDENT THAT FINDS YOU UNPREPARED

Every organization that deploys AI will eventually experience an AI incident. That is not a prediction about risk management failure. It is a statistical certainty about complex systems: models drift, data pipelines corrupt, vendor infrastructure degrades, and edge cases accumulate until they produce outputs that no one designed and no one expected. The question is not whether an AI incident will occur on your watch. The question is whether, when it does, your organization has the protocol to respond in a way that contains the harm, satisfies the governance obligation, and produces a record that demonstrates the response was as deliberate as the governance program that preceded it.

Most organizations do not have that protocol, and the reason is entirely understandable. Incident response protocols are built in response to incidents. The organization that has never experienced a significant AI failure has no institutional memory to draw on, no lessons-learned to incorporate, and no board pressure to prioritize a protocol for a failure that has not yet occurred. Every hour spent building an AI incident response protocol is an hour spent on governance for a problem that does not yet exist, and those hours compete, every day, with governance for problems that do.

> ***Incidents are not governance failures. Unprepared responses to incidents are. The executive who builds the protocol before an incident occurs is the executive whose organization survives one.***

The cost of the unprepared response is not measured only in the harm the incident produces. It is measured in the compounded harm that an uncoordinated response creates on top of the underlying failure. The COO whose opening story begins this chapter faced seven customer complaints. Her uncoordinated four hours of investigation, during which no one could answer the basic questions about scope, accountability, or regulatory obligation, allowed the incident to grow while the response searched for a chain of command. By the time the investigation was properly structured, eleven additional customers had filed complaints and a state regulator had received an inquiry from a consumer advocacy organization. The protocol would not have prevented the model failure. It would have compressed the response from days to hours, bounded the scope before it expanded, and produced a documented response that demonstrated organizational control rather than organizational improvisation.

The seven steps in this chapter are the protocol the COO needed that evening. They are sequenced in the order that incident response must proceed and assigned to the accountability chain links that already exist in your governance structure. If you have completed the governance work in the prior chapters, you already have the roles. This chapter gives those roles their incident instructions.

THE AI INCIDENT RESPONSE PROTOCOL: SEVEN STEPS

The protocol below is organized around three governing principles. First: sequence matters more than speed. The instinct in any incident is to act immediately across all fronts simultaneously. That instinct produces crossed communication, inconsistent information reaching different stakeholders, and decisions made without the scope assessment that should inform them. The seven steps are sequential for a reason. Step 2 (Contain) cannot be executed well without Step 1 (Declare). Step 4 (Communicate) cannot be executed responsibly without Step 3 (Assess). The protocol is a sequence, not a checklist.

Exhibit 7: AI Incident Response Protocol

EXHIBIT 7: AI INCIDENT RESPONSE PROTOCOL

Activate this protocol when: an AI system produces an output that causes or may cause harm to a customer, employee, or the organization; an output deviates materially from expected behavior; or a regulatory inquiry is received involving an AI system.

Step	Phase	Required Actions	Responsible Link	Documentation Required
1	DECLARE	Activate incident status. Notify Technical Owner and Business Owner. Confirm the incident meets declaration threshold. Time-stamp activation. Do not wait for full picture: declare on reasonable belief.	Technical Owner (Link 2) triggers. Business Owner (Link 3) confirms.	Incident declaration record: system name, time of declaration, who declared, basis for declaration. This record is your Day 1 document.

2	CONTAIN	Identify scope of affected outputs or decisions. Suspend, restrict, or isolate the system if safe to do so. If suspension would create greater harm (e.g., life-safety system), document the decision not to suspend.	Technical Owner (Link 2) Business Owner (Link 3)	Scope document: how many outputs, decisions, or individuals affected. Containment action taken and rationale. If not contained: documented reason.
3	ASSESS	Determine magnitude, affected parties, and potential legal or regulatory implications. Notify Risk Officer. Is this a reportable incident under any contract, regulation, or policy? Engage legal counsel at this step.	Risk Officer (Link 4) Legal Counsel	Assessment memo: magnitude, affected parties, legal and regulatory exposure analysis, recommendation on reporting obligations.
4	COMMUNICATE	Execute notification obligations in order: (a) internal: Executive Sponsor and affected business leadership; (b) vendor: trigger vendor notification clause; (c) external: regulatory or affected party notification per legal advice. Do not communicate externally without legal clearance.	Executive Sponsor (Link 5) Legal Counsel	Communication log: who was notified, when, by whom, in what form. Retain all communications as governance records.
5	INVESTIGATE	Conduct root cause analysis with documentation. Did the failure originate in: model drift, vendor retraining, data input quality, governance gap, or operational error? Preserve all system logs and output records before investigation begins.	Technical Owner (Link 2) Risk Officer (Link 4)	Root cause analysis document: technical findings, contributing governance factors, what governance controls did or did not function as designed.
6	REMEDIATE	Restore, retrain, restrict, or decommission the system based on root cause findings. Remediation must include: correction of affected outputs where possible; system fix or replacement; governance policy update if a gap contributed.	Business Owner (Link 3) Technical Owner (Link 2)	Remediation plan and completion record: what was done, when, by whom, and how the fix was verified.
7	DOCUMENT & CLOSE	Build the complete incident audit trail before closing the incident record. Present summary to Executive Sponsor for sign-off. Add to Section 3 (Incident Log) of the next Board AI Governance Summary. Do not close the incident record without Executive Sponsor signature.	Executive Sponsor (Link 5) Board Committee (Link 6) for Tier 3	Closed incident record: all steps documented, root cause confirmed, remediation verified, board summary updated, Executive Sponsor signature. Retained per document retention policy.

Declaration threshold (define before an incident occurs): An incident must be declared when an AI system output causes harm, potential harm, or material deviation from its governed operating parameters. Threshold applies regardless of severity. Severity determines escalation speed, not whether to declare.

Second: the accountability chain is the incident command structure. You do not need a separate incident management team if the accountability chain is populated and functioning. Each link in the chain has a specific role in each step of the protocol, and those roles are assigned below. The incident response protocol is the accountability chain activated under pressure.

Third: documentation is not the last step. It is a continuous obligation from Step 1 forward. Every decision, every communication, every finding, and every action taken during an incident response must be documented as it occurs, not assembled after the incident is resolved. The documentation is the governance record. It is what the regulator will ask for. It is what the board committee will review. It is what distinguishes a governed organization from an organization that improvised its way through a failure.

STEP 1, DECLARE
Establish incident status and activate the accountability chain

WHY THIS STEP EXISTS
An incident that is not formally declared is an incident that is being investigated informally, communicated inconsistently, and managed without a command structure. Declaration is the organizational act that converts a problem being looked into by whoever found it into an incident being managed by the people who are accountable for it. Until declaration occurs, there is no incident command structure, no communication protocol, and no documentation obligation. Declaration is not an escalation of severity. It is the activation of the governance structure that severity will be assessed within.

HOW TO EXECUTE IT
The person who discovers or receives the first credible report of a potential AI incident declares incident status using the organization's defined declaration threshold (specify in the governance policy: any report of an AI output that may have harmed or may harm an individual; any internal finding of model performance degradation below the tier-specific threshold; any vendor notification of a system anomaly). Declaration is communicated to the Technical Owner and the Business Owner simultaneously, in writing, with the timestamp of declaration recorded. The declaration triggers the accountability chain's activation: all links from the declaring party upward are notified per the governance policy escalation timeframes.

RESPONSIBLE LINK
Link 2 (Technical Owner) or Link 3 (Business Owner), depending on who first identifies the incident. Either may declare. Neither may defer declaration pending internal investigation.

MILESTONES
24 HOURS *Declaration made in writing to Technical Owner and Business Owner; escalation chain notification initiated.*

72 HOURS *Risk Officer (Link 4) briefed and incident log opened; scope investigation underway.*

30 DAYS *Declaration and initial scope summary entered in the governance record. If incident remains open at 30 days, board committee notification required regardless of resolution status.*

STEP 2, CONTAIN
Identify scope and isolate the system if warranted

WHY THIS STEP EXISTS
The containment step has one purpose: to prevent the incident from expanding while the assessment step determines what it already is. An AI system that is producing harmful outputs continues to produce them until it is stopped or its outputs are intercepted. An incident whose scope is not bounded continues to expand until it is contained. Containment is not remediation, it is the pause that makes assessment and remediation possible without a moving target.

HOW TO EXECUTE IT
The Technical Owner assesses, within the first four hours of declaration, whether the system should be suspended, throttled, or allowed to continue operation with enhanced human review of outputs while the investigation proceeds. The suspension decision is made by the Business Owner in consultation with the Technical Owner and the Risk Officer. For Tier 3 systems, the Executive Sponsor is notified of the suspension decision before it is implemented. Containment also includes identifying the population of outputs that may have been affected since the earliest plausible onset of the failure: not all incidents have a clear start date, and the Technical Owner must provide an estimated onset range to the Business Owner within 24 hours of declaration.

RESPONSIBLE LINK
Link 2 (Technical Owner) assesses and recommends. Link 3 (Business Owner) decides. Link 5 (Executive Sponsor) approves suspension of Tier 3 systems.

MILESTONES
24 HOURS *Suspension or enhanced-review decision made and implemented; estimated affected output population identified.*
72 HOURS *Affected individual or entity count established to the extent determinable from available data.*
30 DAYS *Containment measures maintained or lifted based on investigation findings; system status documented.*

STEP 3, ASSESS
Determine magnitude, affected parties, and regulatory triggers

WHY THIS STEP EXISTS
The assessment step is where the organization learns what it is dealing with. The declaration and containment steps operate on incomplete information, they activate the protocol before the full picture is known because waiting for the full picture allows the incident to expand. The assessment step is the structured effort to develop that picture as completely as possible within the 72-hour window, so that the communication and investigation steps are informed by accurate scope rather than initial estimates.

HOW TO EXECUTE IT

The Risk Officer leads the assessment, coordinating inputs from the Technical Owner (system-side data), the Business Owner (operational impact data), Legal (regulatory trigger analysis), and the vendor (vendor-side data). The assessment produces four outputs: the affected party count and identity (individuals, entities, or categories affected by the incident); the harm magnitude (actual harm experienced, potential harm identified, harm that was avoided by containment); the regulatory trigger analysis (which regulatory notification obligations are triggered by the incident, to which regulators, and within what timeframes); and the severity classification (minor: no external harm, no regulatory trigger; moderate: potential external harm, regulatory notification likely; major: confirmed external harm, multiple regulatory obligations, board notification required).

RESPONSIBLE LINK

Link 4 (Risk Officer) leads and coordinates. Link 6 (Board Committee Chair) is notified directly if severity classification is Major.

MILESTONES

24 HOURS *Preliminary assessment complete: affected party count estimated, harm magnitude characterized, regulatory triggers identified.*

72 HOURS *Full assessment documented: severity classification confirmed, regulatory notification obligations confirmed with Legal.*

30 DAYS *Assessment findings incorporated into the incident investigation report.*

STEP 4, COMMUNICATE

Execute internal escalation and external notification protocol

WHY THIS STEP EXISTS

The communication step is where most organizations make their most consequential errors. Communication that is too slow allows harm to expand and regulators to learn about the incident from sources other than the organization. Communication that is too fast, before scope is known and messaging is coordinated, produces inconsistent statements to different stakeholders that become the incident's second problem. The communication step requires a specific sequence: internal before external; executive sponsor before board; legal review before regulatory notification; affected parties before public disclosure if disclosure is required.

HOW TO EXECUTE IT

Internal escalation: Risk Officer notifies Executive Sponsor with assessment findings. Executive Sponsor notifies Board Committee Chair if severity is Moderate or Major. For Major incidents, the full board committee is notified before any external communication. External notification: Legal determines whether affected individuals, regulatory bodies, or the public require notification, in what form, within what timeframe, and with what content. For each external communication, a draft is reviewed by Legal and the Executive Sponsor before transmission. All external communications are logged with timestamp, recipient, content, and sender. The vendor is notified of the incident under the incident disclosure provisions of the vendor contract (Non-Negotiable 4 from Chapter Twelve) and asked to confirm whether the incident has a vendor-side cause requiring vendor-side disclosure obligations.

RESPONSIBLE LINK

Link 5 (Executive Sponsor) approves all external communications. Link 4 (Risk Officer) manages internal escalation. Legal owns regulatory notification content and timing.

MILESTONES

24 HOURS *Internal escalation complete; Executive Sponsor briefed; Board Committee Chair notified if Moderate or Major.*

72 HOURS *All required regulatory notifications filed; affected individuals notified if required; vendor informed and vendor-side obligations confirmed.*

30 DAYS *All communication log entries verified complete; board committee has received written incident summary.*

STEP 5, INVESTIGATE

Conduct root cause analysis with documented timeline

WHY THIS STEP EXISTS

The investigation step is not the same as the assessment step. Assessment answers the question of what happened and to whom. Investigation answers the question of why it happened and who or what caused it. The distinction matters because remediation. Step 6, must address the root cause rather than the symptom. An organization that remediates a bias output by removing the system's access to the data field that produced it, without understanding why the model weighted that field as it did, has addressed a symptom. The root cause investigation is what produces durable remediation rather than recurrent incidents.

HOW TO EXECUTE IT

The Technical Owner leads the root cause investigation, working with the vendor if the cause may be vendor-side. The investigation produces a formal timeline: the estimated onset of the failure; the detection event (what surfaced the failure and through what mechanism); the period of undetected operation (how long the failure operated before detection); the scope of affected outputs during that period; and the root cause finding (model drift, data corruption, training data issue, vendor retraining, human override rate change, or other). The root cause finding is reviewed by the Risk Officer and Legal before it is finalized, because root cause findings have implications for regulatory disclosure, liability assessment, and remediation design.

RESPONSIBLE LINK

Link 2 (Technical Owner) leads. Link 1 (Vendor) cooperates under regulatory cooperation provision of the vendor contract. Link 4 (Risk Officer) reviews findings before finalization.

MILESTONES

24 HOURS *Preliminary timeline established; vendor cooperation requested if vendor-side cause is indicated.*

72 HOURS *Root cause finding drafted and under Legal review.*

30 DAYS *Root cause finding finalized, reviewed by Executive Sponsor, and incorporated into the incident report.*

STEP 6, REMEDIATE

Restore, retrain, or decommission, with governance approval

WHY THIS STEP EXISTS

The remediation step is where the organization acts on the root cause finding. The options are restoration (the system is restored to operation after the issue is corrected and tested), retraining (the model is retrained to address a training-data root cause, with governance review of the retrained model before restoration), and decommissioning (the system is removed from operation because the root cause cannot be remediated or the risk profile of continued operation is not justified by the system's value). The remediation decision is a governance decision, not a technical one. The Technical Owner recommends; the governance structure decides.

HOW TO EXECUTE IT

The Technical Owner produces a remediation recommendation with three elements: the proposed remediation action (restore, retrain, or decommission); the evidence that the proposed action addresses the root cause; and a post-remediation testing protocol that will confirm the issue is resolved before the system returns to full operation. For Tier 2 and Tier 3 systems, the remediation recommendation is reviewed and approved by the Risk Officer and the Executive Sponsor before implementation. For Tier 3 systems, the board committee is informed of the remediation decision at or before the next quarterly meeting. If the remediation involves retraining by the vendor, the retraining notification clause (Non-Negotiable 2 from Chapter Twelve) applies, and the post-retraining review protocol is activated.

RESPONSIBLE LINK

Link 2 (Technical Owner) recommends. Link 4 (Risk Officer) and Link 5 (Executive Sponsor) approve for Tier 2 and Tier 3. Link 6 (Board Committee) is informed for Tier 3.

MILESTONES

24 HOURS *Remediation recommendation approved by appropriate governance authorities.*

72 HOURS *Remediation implemented and post-remediation testing completed; system restored, retrained, or decommissioned.*

30 DAYS *Post-remediation monitoring confirms root cause is resolved; any retraining review documented and closed.*

STEP 7, DOCUMENT

Build the audit trail before closing the incident

WHY THIS STEP EXISTS

An incident that is resolved without a complete governance record is an incident that produced harm, consumed organizational resources, and generated no institutional learning. The documentation step converts the seven-step response into a permanent governance record that serves four purposes: regulatory evidence (the documented response is what regulators ask for when they examine an incident); board reporting (the incident log in the governance summary Section 3 requires this documentation); institutional learning (the root cause finding and remediation decision become the evidence base for governance improvements); and legal protection (a documented response is evidence of the reasoned, deliberate decision-making that the Business Judgment Rule protects; an undocumented response is not).

HOW TO EXECUTE IT

Before the incident is formally closed, the Risk Officer confirms that the incident record is complete: declaration timestamp and basis; containment decision and rationale; assessment findings (affected parties, harm magnitude, severity classification); communication log (every internal and external communication, by recipient, timestamp, and content); root cause finding (as reviewed and finalized); remediation decision and approval record; post-remediation testing results; and a lessons-learned summary (what governance gap, if any, contributed to the incident, and what governance change has been or will be made to reduce recurrence). The completed incident record is submitted to the Executive Sponsor for review and then added to the governance record. The incident log entry in the Board AI Governance Summary is updated. The incident is formally closed by the Executive Sponsor in writing.

RESPONSIBLE LINK

Link 4 (Risk Officer) confirms completeness. Link 5 (Executive Sponsor) reviews and formally closes. Incident record enters the governance documentation.

MILESTONES

24 HOURS *All active documentation current; no open communication obligations.*

72 HOURS *Incident record substantially complete pending post-remediation testing.*

30 DAYS *Incident formally closed by Executive Sponsor; board governance summary updated; lessons-learned summary completed and distributed.*

THE GOVERNANCE FAILURE WITHIN THE INCIDENT

The most consequential governance finding from most AI incidents is not the incident itself. It is the discovery that the governance infrastructure that should have detected the incident earlier, override rate monitoring, output review cadence, vendor notification clause, accountability chain coverage, was absent, incomplete, or not functioning as designed. When the root cause investigation reveals a governance gap rather than (or in addition to) a technical failure, that finding must be documented in the incident record and presented to the board committee as a governance finding requiring a governance response. An incident that reveals a governance gap that is then quietly closed without board visibility has not produced institutional learning. It has produced a concealed governance failure on top of an incident.

The board committee that learns about a governance gap from a regulator rather than from its own Executive Sponsor has not been governed. It has been managed.

BUILD THE PROTOCOL BEFORE YOU NEED IT

A protocol deployed for the first time during a live incident is the most expensive version of governance you can buy. The organization that has never exercised it will spend the incident learning it. The tabletop exercise is the investment that converts the protocol from a document into a capability, and it costs less than the first four hours of an uncoordinated response.

ELEVEN DAYS TO GOVERNED

The COO who opened this chapter called me again eleven days after the customer complaint that had found her without a protocol. She had spent those eleven days building one, activating it retroactively, and running the benefits determination system incident through all seven steps simultaneously rather than in sequence, which is what happens when you build the protocol during the incident rather than before it. The experience had cost her organization more than the incident itself would have.

The incident itself affected thirty-one customers whose benefits determinations had been influenced by a model drift that had begun six weeks before the first complaint. The drift had been detectable from the override rate data, the rate had declined from 28 percent to 9 percent over a four-month period, crossing the Tier 2 reclassification threshold without triggering a governance review because there was no automated monitoring and the quarterly review had been deferred. The root cause was a vendor data update that had introduced a weighting change in one demographic category. The vendor had not issued a notification because their standard terms did not require one.

The harm to the thirty-one customers was real but bounded: incorrect determinations had been made, and correcting them required individual case reviews that took an average of ninety minutes each. The compliance team that conducted the reviews worked the corrections over six business days. Forty-seven hours of corrective casework, with one formal regulatory notification to the state insurance commissioner and one individual notification to each affected customer.

The organizational cost of the uncoordinated response added significantly to that baseline. The first four hours of searching for the accountable owner, during which seven additional customers filed complaints that could have been intercepted by a declared incident management structure. The three days before the vendor confirmed the data update, during which the Technical Owner could not close the root cause investigation because the vendor cooperation clause did not exist in the original contract. The two cycles of external communication that had to be corrected because draft messaging had been sent to one stakeholder group before Legal had reviewed it. The board committee meeting at which the COO had to disclose not only the incident but the governance gaps it had revealed, the deferred quarterly review, the missing vendor notification clause, the override rate decline that had not been monitored, in a room where the committee had previously been told the governance program was functioning.

She told the board committee the truth, completely, including the governance gaps. The committee chair's response was the one the COO had both feared and needed: "Thank you for the full disclosure. I want to see the corrective governance actions documented and confirmed at the next meeting. The incident is resolved. The governance gaps are not yet."

In the six months that followed, the COO built the protocol, activated the override rate monitoring for all Tier 2 and Tier 3 systems, amended the vendor contract with all five non-negotiables, and conducted the first tabletop exercise for the benefits determination system. At the next board committee meeting, she presented the corrected governance documentation alongside the incident's closed record. The committee chair said: "This is what a governed organization looks like after a failure. The incident was a governance cost. The response was a governance investment."

The COO told me that she wished she had built the protocol before the incident, but she was grateful she had built it after rather than not at all. What she had done after was more expensive than what she would have done before. What she had done was still governance.

BUILDING THE PROTOCOL BEFORE THE INCIDENT

The protocol is built in three actions. Each action takes less time than the first four hours of an uncoordinated incident response.

1. **Draft the incident declaration threshold and add it to the governance policy today.**

The protocol activates on declaration, and declaration requires a defined threshold. Without a written threshold, the declaration decision is a judgment call made by whoever discovers the incident under pressure. Write one paragraph that defines the conditions that trigger incident declaration for your organization: any credible report of an AI output that may have harmed an individual; any internal finding of model performance degradation below a defined threshold; any vendor notification of a material system anomaly; any regulatory inquiry that references a specific AI system's outputs. The threshold does not need to be exhaustive. It needs to be clear enough that the Technical Owner and Business Owner, finding something that matches any of the listed conditions, will declare rather than investigate informally. Add the threshold paragraph to the AI governance policy before the week ends.

Write the threshold broadly enough to capture incident types you have not anticipated. A threshold that says "declare when a system produces discriminatory outputs" fails to capture a data breach, a vendor model substitution, or a regulatory inquiry. The declaration threshold should read: any AI system output that causes harm, potential harm, or material deviation from governed operating parameters. Broad coverage means you declare more frequently and close more incidents. Narrow coverage means you miss the incidents your threshold did not name. Success looks like a written declaration threshold your Legal team has reviewed and your accountability chain owners all understand before an incident occurs.

2. **Assign incident response roles to each accountability chain link in the governance documentation.**

The protocol's seven steps have defined responsible links. For each Tier 2 and Tier 3 system in your asset register, document who holds each link and what their incident response obligation is. The documentation does not need to be elaborate, a single page per system that maps each of the seven protocol steps to the named link holder for that system is sufficient. The page lives in the governance documentation folder alongside the accountability chain record. When an incident occurs, the Technical Owner's first action after declaration is to pull that page, confirm the link holders are reachable, and initiate notification. The protocol page is the incident command structure in a form anyone can follow under pressure.

Most common mistake: completing the role assignment for Tier 3 systems only. Tier 2 incidents are more frequent than Tier 3 incidents precisely because Tier 2 systems are more numerous and less intensively monitored. A Tier 2 system without a documented incident response role assignment is a system whose incident will be managed by improvisation rather than protocol. Complete the role assignment for every Tier 2 and Tier 3 system. Success looks like a completed one-page incident response role sheet for every Tier 2 and Tier 3 system, filed in the governance documentation and confirmed with each named link holder before the next quarterly governance review.

3. **Schedule and conduct a tabletop exercise for your highest-consequence Tier 3 system before the next quarterly board meeting.**

The tabletop exercise is a ninety-minute structured simulation in which the accountability chain for a specific Tier 3 system walks through a hypothetical incident scenario using the seven-step protocol. The scenario does not need to be elaborate, a specific output failure with a defined scope and a vendor-side root cause is sufficient. The exercise tests whether the protocol is understood, whether the declaration threshold is clear, whether the communication sequence is workable, and whether the documentation obligation is taken seriously. The tabletop findings are documented and added to the governance record. The board committee is notified that the exercise occurred, with a summary of findings, at the next quarterly governance summary presentation.

Design the tabletop scenario for maximum discomfort, not maximum clarity. The scenario that is easily navigated by your accountability chain confirms that the chain works in comfortable conditions. The scenario that exposes a gap in the declaration threshold, an ambiguous escalation between Link 3 and Link 4, or a vendor communication failure, is the scenario that makes the protocol stronger before it matters. Success looks like a tabletop that ends with at least three protocol improvements documented and assigned to named owners for completion before the next exercise.

THE PROTOCOL THAT EARNS THE RIGHT TO FAIL

Every organization that governs AI will experience an AI incident. The governance literature does not say this often enough, perhaps because it sounds like an admission of inadequacy. It is the opposite. It is the acknowledgment that complex systems operating in complex environments will occasionally produce unexpected outcomes, and that the measure of a governed organization is not whether it avoids all failures but whether it responds to them in a way that bounds the harm, meets the governance obligation, and produces a record that demonstrates the response was as deliberate as the program that preceded it.

The COO in this chapter's story built her protocol after the incident. She governed the aftermath with the tools she built during it. The committee chair's observation, "the incident was a governance cost; the response was a governance investment", holds whether the protocol is built before or after the failure. But an investment made before the failure costs less and produces more. The protocol built before the incident is the one that compresses days into hours, prevents compounding harm, and generates board confidence rather than board disclosure. That is the protocol this chapter gives you.

Incidents are not governance failures. Unprepared responses to incidents are. The executive who builds the protocol before the incident occurs is the one whose organization can survive a failure with its credibility intact, because the protocol is already the answer to the question the regulator, the board, and the affected party will all ask: what did you do, and in what order?

Chapter Twelve addresses the governance layer that most executives discover too late: the vendor contract. When an incident occurs and the accountability chain reaches Link 1, what happens next depends entirely on what is in the contract with the vendor whose system failed. That contract was probably negotiated before governance was on the agenda. It is not too late to change it.

> ***The protocol built before the incident is the one that converts a failure into a governance record. Build it before you need it. The incident will test the protocol. The protocol will protect the organization.***

BOARDROOM TAKEAWAY | *The Vendor Negotiation No One Prepares For*

What a board member should be able to ask:

- Do our vendor contracts for Tier 2 and Tier 3 AI systems contain all five non-negotiable provisions?
- Which vendor contracts require amendment before renewal, based on the non-negotiables audit?

What a CFO or risk leader should have documented:

- A completed non-negotiables audit for every Tier 2 and Tier 3 vendor contract
- A priority list of contracts requiring amendment, ordered by renewal date and risk tier

What evidence should exist:

- At least one vendor contract amended to include the retraining notification provision
- A documented contract review process for all future AI vendor agreements that applies the five non-negotiables as standard

CHAPTER TWELVE: PART III: THE EXECUTION

The Vendor Negotiation No One Prepares For

Five Contract Gaps That Will Cost You When They Matter Most

The vendor's attorney said: "I'm sorry, but that's not in our agreement."

> **Illustrative Scenario:** *The following scenario is representative of documented patterns in executive AI governance. Names, organizations, and specific details are illustrative.*

The General Counsel had been on the call for eleven minutes. She had expected a cooperative conversation about the data disclosure the regulatory examination required. She had a vendor whose AI system had produced the outputs at issue, a regulator who needed documentation of what data the system had processed and how the model had been trained, and a contract that said nothing useful about either.

She asked about the incident disclosure clause. The vendor's attorney said there was no incident disclosure clause. She asked about the model training documentation. The attorney said that was proprietary. She asked whether the vendor was obligated to cooperate with regulatory inquiries. The attorney said the vendor would cooperate voluntarily with reasonable requests but had no contractual obligation to do so. She asked what "reasonable" meant. The attorney said that would need to be assessed case by case.

She put the call on hold and called the CFO. She said: "We have a regulatory examination, a vendor whose system is under examination, and a contract that gives us nothing. I need you to understand that every governance document we've built is useful except the one we needed most, which was a contract with this vendor that required them to cooperate when we need them to."

The examination took four months. The vendor cooperated, eventually, voluntarily, and on its own terms. The organization produced the required documentation. No enforcement action resulted. The cost of the four months was not the regulatory outcome. It was the eleven months of renegotiation that followed, during which the organization's legal team rewrote every AI vendor contract in the portfolio with the five provisions it had learned, expensively, were missing.

This chapter gives you those five provisions before the examination arrives.

THE CONTRACT THAT GOVERNS NOTHING

AI vendor contracts are negotiated by procurement professionals and technology leaders who are expert in the dimensions of vendor contracting they have always managed: price, service levels, intellectual property, liability limits, and termination rights. Those dimensions matter. They are not the dimensions that govern AI.

The dimensions that govern AI, model performance accountability, retraining disclosure, data ownership on termination, incident disclosure timelines, and regulatory cooperation obligations, are not standard provisions in commercial technology contracts. They do not appear in vendor templates. They are not requested by procurement functions that have not encountered them before. They are not offered by vendors whose legal teams have drafted contracts specifically to avoid them. The result is a portfolio of AI vendor contracts that are commercially complete and governance-empty: they specify what the vendor will deliver, what it will charge, and what happens if either party wants to leave. They do not specify what the vendor owes the organization when the model they have sold fails, changes, or attracts regulatory attention.

> ***Your AI vendor is not your partner in governance. They are a counterparty. The contract that treats them accordingly is not adversarial. It is accountable.***

The reason most organizations have not closed these gaps is not that they lack negotiating leverage. It is that they did not know what to ask for. The procurement team that negotiates a SaaS contract for an HR platform has a well-developed set of requirements built through years of contracting experience, industry standards, and institutional learning. The procurement team that negotiates a contract for an autonomous credit-scoring model is working in a domain where industry standards are still forming, regulatory requirements are still emerging, and the organizational learning about what the contract needs to contain has not yet accumulated.

The five provisions in this chapter are what organizations discover they needed after the governance failure. They are synthesized here so that every organization can require them before signing.

The cost of the missing provisions is not theoretical. It is the four months of regulatory examination in this chapter's opening story. It is the six-week service failure in Chapter Seven's logistics company story, which a retraining notification clause would have compressed to days. It is the eleven-week benefits administrator failure from Chapter Seven, which a performance accountability clause would have surfaced months earlier. In each case, the organization's governance infrastructure was adequate at every layer except the one closest to the AI system itself: the contract that governed what the vendor owed when something went wrong.

THE FIVE AI CONTRACT NON-NEGOTIABLES

Each of the five provisions below is a governance requirement dressed in contract language. The model language provided for each is a starting point, not a finished clause, your General Counsel will adapt it to your organization's contracting conventions and the specific system being governed. What the model language provides is the governance intent: the specific obligation the provision creates, the specific trigger that activates it, and the specific remedy that enforces it. Without those three elements, the provision is a statement of hope rather than a contractual obligation.

These provisions apply to every AI vendor contract covering a Tier 2 or Tier 3 system. For Tier 1 systems, provisions 3 (data ownership) and 4 (incident disclosure) are the minimum standard. The negotiation posture for each provision is addressed below, including the pushback you will receive and how to hold the line. Vendors who refuse all five provisions after genuine negotiation are vendors whose systems should be tiered at Tier 3 by default, regardless of their designed decision authority level, because a vendor who will not accept governance obligations is a vendor you cannot govern.

NON-NEGOTIABLE 1, PERFORMANCE ACCOUNTABILITY CLAUSE

WHY IT MATTERS

AI models degrade. They drift. They produce outputs that were within acceptable performance parameters at deployment and outside them eighteen months later, without anyone necessarily noticing until a failure reveals the drift. A performance accountability clause creates a contractual standard against which degradation can be measured and a defined remedy when the standard is not met. Without it, the vendor's only obligation is to provide the service, not to ensure the service performs at the level that justified deploying it.

WHAT IS TYPICALLY MISSING

Standard SaaS contracts define uptime and availability obligations but not model performance obligations. The vendor commits to keeping the system running; they do not commit to ensuring the model produces accurate outputs within a defined tolerance, or to disclosing when accuracy has degraded beyond that tolerance. The organization discovers degradation when a human reviewer notices something wrong or when an audit reveals a pattern, not because the vendor reported it.

MODEL LANGUAGE

Vendor warrants that the AI model ('System') will maintain performance metrics within the thresholds specified in Exhibit [X] ('Performance Standards') as measured by the assessment methodology described therein. Vendor shall notify Customer within [10] business days if internal monitoring indicates the System's performance has degraded below any Performance Standard threshold. Customer shall have the right to commission an independent performance assessment at any time upon [30] days' notice; if the assessment confirms degradation below the Performance Standards, Vendor shall remediate at no charge within [60] days or Customer may terminate without penalty.

COMMON VENDOR PUSHBACK

"Our performance varies based on the quality of your input data, so we can't guarantee specific accuracy thresholds."

HOW TO HOLD THE LINE

Respond with specificity: the Performance Standards exhibit should reflect the accuracy and fairness metrics that were used to justify the system's deployment and tier classification. Those metrics already exist in your governance documentation. You are not asking the vendor to meet a new standard. You are asking them to meet the standard they represented when you signed. A vendor who claims they cannot guarantee the performance that justified the original contract is making a statement about their confidence in their own product. That statement belongs in your tier classification decision.

NON-NEGOTIABLE 2, RETRAINING NOTIFICATION CLAUSE

WHY IT MATTERS

A model that is retrained is, in a governance-relevant sense, a different model. Its outputs may change. Its bias profile may shift. Its performance in specific demographic segments or geographic clusters may improve or deteriorate. The organization that deployed a model subject to governance approval has not approved the retrained version. Without a retraining notification clause, the vendor can change the model the organization is running without triggering the governance review that a materially different model should require. The logistics company in Chapter Seven lost six weeks of service quality to a retraining it did not know had occurred. The clause prevents exactly that.

WHAT IS TYPICALLY MISSING

Vendors routinely update and retrain models as part of standard product improvement. Standard SaaS terms typically reserve the right to update the service without notice, and those terms cover model retraining unless specifically carved out. The organization discovers a retraining has occurred when it notices output changes, or, as in Chapter Seven, when an external investigation surfaces it.

MODEL LANGUAGE

Vendor shall provide Customer with written notice no less than [5] business days prior to any planned retraining of the System model, including a description of the training data to be used, the anticipated changes to model behavior, and the expected effective date. For emergency retraining required to address a critical performance failure, Vendor shall provide notice within [24] hours of retraining completion. Customer shall have the right to conduct a post-retraining review of System outputs for a period of [30] days; if the review identifies material changes to System outputs that affect Customer's governance obligations, Customer may require Vendor to roll back to the prior model version pending Customer's governance review.

COMMON VENDOR PUSHBACK

"Our model improvement process is continuous and we can't always predict when retraining will occur or what specific changes will result."

HOW TO HOLD THE LINE

Distinguish between continuous learning (acceptable to notify at defined intervals, such as monthly) and material retraining events (requiring advance notice). Most vendors can accommodate this distinction. A vendor who genuinely cannot predict when their model will be retrained is a vendor running a model they do not adequately monitor, which is itself a governance signal worth naming in the negotiation.

NON-NEGOTIABLE 3, DATA OWNERSHIP AND PORTABILITY CLAUSE

WHY IT MATTERS

When an organization terminates an AI vendor relationship, two questions must be answered immediately: what happens to the data the vendor processed, and can the organization retrieve its data in a usable format? The first question is a privacy, security, and regulatory obligation. The second is an operational and governance continuity obligation. Standard termination provisions address service cessation; they do not reliably address data destruction certification, data retrieval format, or the vendor's right to retain aggregated or anonymized versions of the organization's data for model improvement purposes after termination.

WHAT IS TYPICALLY MISSING

Many AI vendor contracts include data rights language that grants the vendor the right to use 'aggregated or anonymized' customer data to improve their models. This language is often in the standard terms and conditions rather than the main agreement body, which means it is frequently unread and unreviewed by the legal team that negotiates the principal contract. Post-termination data rights, data retrieval formats, and destruction certification timelines are typically absent entirely.

MODEL LANGUAGE

> *All data provided by Customer to Vendor for processing by the System ('Customer Data') remains the sole property of Customer. Vendor shall not use Customer Data for any purpose other than providing the contracted services, including but not limited to model training, benchmarking, or product improvement, without Customer's prior written consent. Upon termination or expiration of this Agreement, Vendor shall: (i) within [30] days provide Customer with a complete export of Customer Data in a mutually agreed machine-readable format; (ii) within [60] days certify in writing that all Customer Data has been deleted from Vendor's systems and any subprocessors' systems; and (iii) cease all processing of Customer Data as of the termination effective date.*

COMMON VENDOR PUSHBACK

"We use aggregated, anonymized data to improve our models for all customers, this benefits you as well."

HOW TO HOLD THE LINE

Aggregation and anonymization do not eliminate the governance concern. If the vendor's improved model, trained partly on your data, is sold to your competitors, you have contributed to a competitive disadvantage without consent. Require explicit opt-in for any use of your data beyond service delivery, and require the vendor to identify by category what data they retain post-termination and under what legal basis. If they cannot answer that question, the data rights clause is more important than they are representing.

NON-NEGOTIABLE 4, INCIDENT DISCLOSURE OBLIGATION

WHY IT MATTERS

An AI vendor incident, a model failure, a data breach, a security compromise, a discovered bias pattern, a regulatory inquiry directed at the vendor's model, is an incident that affects the organization using the model, regardless of whether it originated with the vendor. The organization's accountability chain and incident response protocol assume that it will know about vendor-side incidents quickly enough to respond within the escalation timeframes the governance policy specifies. Without a contractual disclosure obligation, the organization learns about vendor incidents when the vendor chooses to disclose them, which is typically after the vendor has assessed its own exposure and prepared its own response, at a timeline that serves the vendor's interests rather than the organization's governance obligations.

WHAT IS TYPICALLY MISSING

Standard technology contracts require disclosure of data security incidents within legally mandated timeframes (typically 72 hours under GDPR; 30 days under most state breach notification laws). AI governance incidents, model failures, bias findings, performance degradations, regulatory inquiries directed at the model itself, are not data security incidents and are typically not covered by breach notification provisions. They exist in a contractual gap that most organizations do not discover until an incident occupies it.

MODEL LANGUAGE

Vendor shall notify Customer within [48] hours of becoming aware of any of the following: (i) a material failure or unexpected behavior of the System that may affect the accuracy, fairness, or reliability of its outputs; (ii) any data security incident affecting Customer Data; (iii) any regulatory inquiry, investigation, or examination directed at Vendor's AI model or practices that may affect Customer's use of the System; (iv) any material finding from Vendor's internal model monitoring that affects the System's performance characteristics. Notice shall include a description of the incident, Vendor's current understanding of its cause and scope, and Vendor's intended response timeline. Vendor shall provide Customer with status updates no less frequently than every [5] business days until the incident is resolved.

COMMON VENDOR PUSHBACK

"We have our own incident response process and will notify you when we have something definitive to share."

HOW TO HOLD THE LINE

The governance reason for early notification is precisely that you do not yet have something definitive. Your accountability chain and board notification protocols run on timelines that cannot wait for the vendor's internal assessment to be complete. The 48-hour notification requirement is not a request for a completed investigation, it is a requirement for an initial disclosure that allows your governance process to begin. Frame the negotiation around your regulatory obligations, not your preference: 'Our regulator expects us to notify them promptly of incidents affecting our AI systems. We can only do that if our vendor notifies us promptly.'

NON-NEGOTIABLE 5, REGULATORY COOPERATION CLAUSE

WHY IT MATTERS

When a regulator examines an AI system, the examination covers the system's design, training, outputs, and governance, much of which is held by the vendor, not the organization. Without a contractual cooperation obligation, the vendor's participation in the regulatory response is voluntary, and voluntary cooperation is governed by the vendor's own legal exposure assessment rather than the organization's governance obligations. The General Counsel in this chapter's opening story spent four months negotiating voluntary cooperation that a regulatory cooperation clause would have made automatic. That four months was not an anomaly. It is the standard experience of organizations that discover this gap under examination conditions.

WHAT IS TYPICALLY MISSING

Standard vendor contracts include an indemnification structure that governs liability between the parties but does not specify the vendor's obligation to cooperate with third-party regulatory processes. The vendor's posture in a regulatory examination is governed by its own legal counsel's assessment of its exposure, an assessment that may be adversarial to the organization's interests, even when the organization and vendor are not in direct dispute with each other.

MODEL LANGUAGE

Vendor shall, upon Customer's reasonable request, cooperate with any regulatory examination, audit, or investigation involving Customer's use of the System, including by: (i) providing documentation of the System's model architecture, training data sources, and performance assessment methodology to the extent necessary to respond to regulatory inquiries; (ii) making available a qualified technical representative to respond to examiner questions about the System at a mutually agreed time; (iii) providing Customer with all documentation in Vendor's possession that Customer's legal counsel reasonably determines is responsive to the regulatory inquiry. Vendor's obligations under this section are subject to any legal privilege or third-party confidentiality restrictions that Vendor identifies in writing within [10] business days of Customer's request.

COMMON VENDOR PUSHBACK

"We're happy to cooperate voluntarily, but we can't commit contractually to regulatory cooperation because we don't know what regulators might ask for."

HOW TO HOLD THE LINE

The regulatory cooperation clause does not commit the vendor to producing everything a regulator might ever request. It commits the vendor to cooperating with reasonable requests, subject to privilege and third-party confidentiality. That carve-out is already in the model language. A vendor who refuses a reasonable cooperation obligation with a privilege carve-out is signaling that they expect their interests and yours to diverge under regulatory pressure. That signal should affect your tier classification and your procurement decision.

⚠ NEGOTIATION WARNING

The most dangerous moment in AI vendor negotiation is the moment the vendor offers a 'side letter' or 'addendum' that addresses governance provisions outside the main agreement. Side letters are frequently not included in contract management systems, are not reliably surfaced during due diligence, and may not be enforceable in jurisdictions where the main agreement specifies it as the entire agreement. Require all governance provisions to be incorporated into the main agreement or a formally executed amendment. A governance obligation that exists only in a side letter is a governance obligation that will be difficult to enforce when you need it most.

THE AMENDMENT THAT CHANGED THE RELATIONSHIP

The VP of Operations from Chapter Seven, the one whose logistics routing system had failed silently for six weeks after a vendor retraining, returned to the vendor negotiation table three weeks after the failure was resolved. She had the accountability chain documentation, the incident timeline, and a drafted amendment covering all five non-negotiables. She also had something she had not had in the original negotiation: a specific commercial consequence to name.

The vendor's account team initially responded the way account teams typically respond to governance amendment requests: they acknowledged the documentation, said they would need to involve their legal team, and suggested a timeline of six to eight weeks for review. The VP of Operations said: "I'd like to accelerate that. We have a contract renewal in ninety days and my board has asked me to confirm that our AI vendor contracts meet our governance standards before renewal. If we can't get to agreement on these five provisions, I need to begin evaluating alternatives with enough time to make a transition."

The vendor's legal team responded within two weeks.

The negotiation was not frictionless. The performance accountability clause required three rounds of language negotiation before the vendor accepted a formulation that defined performance standards by reference to the metrics in the governance exhibit rather than by absolute thresholds the vendor found objectionable. The retraining notification clause was accepted with a modification: planned retraining would receive five business days' notice, but the vendor added language distinguishing between material retraining events and routine model optimization, with notice required only for the former. The VP of Operations accepted that distinction because it preserved the governance intent, she would know about meaningful model changes, without creating an operational burden that would cause the vendor to underreport.

The data ownership clause and the incident disclosure obligation were accepted with minimal revision. The regulatory cooperation clause required the most negotiation: the vendor's legal team initially proposed language limiting cooperation to "commercially reasonable efforts," which the VP of Operations and her General Counsel rejected. The final language committed to cooperation with reasonable requests, subject to a privilege carve-out with a ten-business-day response requirement to identify any privileges being asserted. That formulation gave the organization what it needed, a commitment to cooperate within a defined timeline, while giving the vendor a legitimate mechanism to protect genuinely privileged information.

The amendment was signed forty-one days after the initial request. Eight months later, the vendor notified the organization of a retraining event per the amended contract. The Technical Owner reviewed the notification, determined that the retraining affected the geographic cluster performance characteristics, and conducted a thirty-day post-retraining output review. The review found no adverse change. The finding was documented in the governance record. The entire governance cycle took less than three weeks and produced documentation that the organization had exercised oversight of a vendor-side model change.

It was, the VP of Operations noted, exactly what six weeks of silent service failure had been the absence of.

The vendor relationship did not become adversarial after the amendment.

It became clearer.

Both parties knew what they owed each other. The governance obligations were not an imposition on the relationship, they were the structure that made a high-consequence vendor relationship professionally manageable rather than dependent on goodwill and informal cooperation. The vendor account team, the VP of Operations noted, seemed relieved to have the governance expectations defined explicitly. They knew what they needed to do. They did it.

CLOSING THE CONTRACT GAPS THIS QUARTER

The five non-negotiables apply to every existing AI vendor contract in your portfolio and every future contract you will sign. Four actions that close the gaps.

1. **The Link 1 contract audit is the starting action. Pull the executed contract for every Tier 2 and Tier 3 AI system in your register. Read it specifically for the five provisions. Where they are absent, mark them for amendment.**

The Link 1 contract audit from Chapter Seven's accountability chain framework is the starting action. Pull the executed contract for every Tier 2 and Tier 3 AI system. Read it specifically for the five provisions: performance accountability, retraining notification, data ownership, incident disclosure, and regulatory cooperation. For each contract, create a simple five-column gap map: which provisions are present, which are partially present, and which are absent. This audit takes approximately thirty minutes per contract and produces the amendment priority list that drives the next three months of vendor negotiation activity.

Include Legal in the contract audit from the beginning. The five provisions have legal implications that a procurement or technology review will not fully surface. Your General Counsel needs to confirm that the provisions, as written in each contract, actually provide the protection they appear to offer. A clause that looks like a performance accountability provision but lacks enforcement teeth is worse than no clause, because it creates the appearance of governance without the substance. Success looks like a contract audit where Legal, Procurement, and the Executive Sponsor review findings together before any vendor conversation begins.

2. **Prioritize by two variables, not one. The first is tier: a Tier 3 system with a missing incident disclosure clause represents a more urgent governance gap than a Tier 1 system with the same absence. The second is contract renewal date: the renewal window is your leverage point. Do not approach a vendor about amendments outside of a renewal discussion unless the gap is critical enough to warrant it.**

Not every gap needs to close this quarter. The amendment priority list should rank contracts by two factors: the system's tier (Tier 3 contracts are highest priority regardless of renewal date) and the contract's renewal or renegotiation window (contracts approaching renewal in the next six months have natural amendment leverage). For Tier 3 systems with contracts more than twelve months from renewal, consider whether the governance gap warrants an out-of-cycle amendment request or whether the risk can be managed through operational protocols until the renewal window. Document the risk management decision for any Tier 3 gap that is not being immediately amended, that documentation is part of the governance record.

Most common mistake: treating all five provisions as equally urgent for all contracts. A Tier 1 system with a missing retraining notification clause represents a different governance priority than a Tier 3 system with a missing regulatory cooperation clause. Allocate legal resources to the highest-consequence gaps first. Success looks like a prioritized amendment list, approved by the General Counsel and the Executive Sponsor, with a target completion date for each contract on the list.

The goal is not perfect contracts. It is governed contracts. A Tier 3 system with three of five provisions in place, with the remaining two on a documented amendment timeline, is a governed vendor relationship. That is where you are headed.

3. **The contract renewal window is the most powerful leverage point in AI vendor negotiation. Vendors who resist governance amendments during the contract term have a material incentive to reach agreement before renewal: the alternative is a competitive process that may not favor them. Six to nine months before any material AI vendor contract renews, initiate the governance amendment conversation explicitly: "As part of our renewal process, we are requiring that all AI vendor contracts meet our governance standards. I want to share the five provisions we need and begin the negotiation with enough time to reach agreement before the renewal date." That framing converts the governance request from a unilateral demand into a mutual precondition for a commercial relationship both parties want to continue.**

Start the governance amendment conversation ninety days before renewal, not at renewal. At the renewal table, both parties have leverage and urgency. Ninety days before renewal, the vendor wants to retain the contract and has time to accommodate reasonable requests without the pressure of an expiring agreement. Governance provisions are easier to add when the vendor is not simultaneously negotiating pricing, term length, and scope. Success looks like an amendment conversation that concludes before renewal discussions begin, because the two conversations have different dynamics and the governance conversation deserves its own.

4. **Add the five provisions to your standard AI procurement checklist for all future contracts.**

Every AI vendor contract your organization signs after today should include the five provisions as a baseline requirement, not an amendment to be negotiated later. Work with your procurement function and

General Counsel to add the five non-negotiables to the standard AI procurement checklist: the list of requirements that must be confirmed before a contract is executed. The model language in this chapter gives procurement and legal the starting point. New contracts that include the provisions from day one negotiate from a posture of governance readiness rather than governance remediation. The vendor who accepts the provisions at signing has demonstrated a governance posture that should inform the system's tier classification. The vendor who refuses all five at signing has demonstrated one that should inform it just as clearly.

Own the procurement checklist as the Executive Sponsor, not as a Legal deliverable. The five provisions reflect governance requirements that you have established for your AI program. Legal confirms that the contract language delivers those requirements. The distinction matters because governance requirements that are delegated entirely to Legal become legal checklists, reviewed at contract time and forgotten between reviews. Governance requirements owned by the Executive Sponsor are active standards, monitored through the accountability chain and updated as the program evolves. Success looks like a procurement checklist signed by the Executive Sponsor and referenced in the AI governance policy, not filed in the legal matter folder.

THE CONTRACT LAYER OF ACCOUNTABLE GOVERNANCE

The General Counsel in this chapter's opening story built a governance program that was thorough at every layer except the one closest to the AI system itself. The asset register named the vendor. The accountability chain assigned a Technical Owner to the vendor relationship. The tier classification appropriately designated the system Tier 3. None of that governance infrastructure could compensate for a contract that gave the vendor no obligations when the regulator arrived.

The five non-negotiables are the contract layer of accountable governance. They do not replace the asset register, the tier model, the accountability chain, or the board summary. They activate those instruments. A performance accountability clause is what makes the tier classification meaningful to the vendor. A retraining notification clause is what makes the Technical Owner's monitoring protocol viable. A regulatory cooperation clause is what makes the accountability chain's Link 1, vendor accountability, a genuine link rather than an aspirational one.

The VP of Operations whose vendor signed the amendment did not have a worse vendor relationship afterward. She had a clearer one. Both parties knew what they owed each other. The five provisions did not impose governance on the relationship, they defined what the relationship was. That definition is what the contract layer of accountable governance produces, and it is what every governance instrument built in the prior eleven chapters requires in order to function.

You have now built the complete governance infrastructure that Part III set out to deliver: the 90-day sprint, the board presentation, the incident protocol, and the vendor contract standards. Part Four turns to what comes next, not the governance of what you have today, but the governance readiness for what is coming. Chapter Thirteen addresses the regulatory landscape that is approaching and how the accountable executive positions their organization to meet it from a position of preparation rather than reaction.

> ***The five non-negotiables give you the leverage to enforce governance standards before a failure. That leverage is not adversarial. It is what the contract layer of accountable governance looks like.***

BOARDROOM TAKEAWAY | *Regulation Is Coming*

What a board member should be able to ask:

- For the regulatory requirements taking effect within eighteen months, where do we currently stand on each of the seven readiness categories?
- Does our governance program, as built, already satisfy the documentation requirements of the frameworks most likely to apply to us?

What a CFO or risk leader should have documented:

- A completed Regulatory Readiness Matrix with honest self-assessments in all seven categories
- A priority remediation list for any category scoring below the defined readiness threshold

What evidence should exist:

- A dated governance readiness assessment that preceded any regulatory examination or inquiry
- Documentation showing that the organization identified and began addressing regulatory gaps proactively

CHAPTER THIRTEEN: PART IV: THE HORIZON

Regulation Is Coming. The Accountable Executive Is Already Ready.

How to Stop Chasing Compliance and Start Leading It

> **Illustrative Scenario:** *The following scenario is representative of documented patterns in executive AI governance. Names, organizations, and specific details are illustrative.*

The examiner's first question was not about the AI. It was about the governance.

"Before we get into the specific systems," she said, placing a notepad on the conference table, "I'd like to understand how your organization governs its AI portfolio. What structure do you have in place?"

The Chief Risk Officer across the table had been preparing for this examination for four months. She had the AI asset register in a binder. She had the TCAIO portfolio summary. She had the Board AI Governance Summary from the prior three quarters. She had the accountability chain documentation for every Tier 2 and Tier 3 system. She had the vendor contract amendments. She had the governance policy, adopted by board resolution, with the date of adoption on the cover page.

She set a single document in front of the examiner: the Regulatory Readiness Matrix. One page. Seven regulatory categories, each mapped against current regulatory status and the organization's governance preparation level, with the specific governance documentation that constituted evidence of preparation cited in each cell.

The examiner reviewed it for three minutes without speaking. Then she said: "This is the most organized regulatory readiness document I have seen from an organization of your size. Walk me through it."

The walkthrough took forty minutes. The examiner asked twenty-two questions. The CRO answered twenty of them from the matrix and its underlying documentation. Two required a follow-up document request. The examination produced no AI governance findings. The CRO told me afterward that she had spent forty-seven hours preparing for the examination. Thirty-eight of those hours had been spent building the governance program over the prior year. Nine had been spent assembling the examination package. The matrix had taken ninety minutes to populate because the governance work it documented already existed.

Thirty-eight hours is the governance investment. Nine hours is the documentation cost. The finding came from the thirty-eight.

THE COMPLIANCE POSTURE THAT COSTS TWICE

Most organizations approach AI regulation the same way they approach every other regulatory development: they wait for sufficient clarity, engage outside counsel to assess their obligations, and implement compliance measures in response to the final regulatory text. This approach is rational for static regulatory regimes with long implementation timelines and clear compliance standards. AI regulation is none of those things. It is dynamic, overlapping across multiple jurisdictions and regulatory bodies, frequently principle-based rather than rule-based, and in several material respects already in force through enforcement actions and examination findings that predate any formal rulemaking.

The cost of the reactive compliance posture is not merely the cost of the compliance measures themselves. It is the cost of building compliance on top of operations rather than into them: the legal review cycles that interrupt ongoing AI deployments, the retroactive governance documentation that has to be created to satisfy examination requests, the board conversations about regulatory exposure that happen after deployments rather than before them, and the vendor renegotiations that are conducted under regulatory pressure rather than commercial leverage. Reactive compliance is governance twice, once when the regulator forces it and once for all the work that should have been done before.

> ***Reactive compliance is governance twice. The proactive executive does not build governance in response to regulation. They discover that the governance they already built is the regulation's answer.***

The reason most organizations have not shifted to a proactive regulatory posture is not that they lack commitment to compliance. It is that the connection between governance work and regulatory readiness has not been made explicit. The Chief Risk Officer who built an AI asset register did not think of it as regulatory preparation. The CFO who required TCAIO analyses for new AI deployments did not frame it as regulatory documentation. The board committee that adopted the AI governance summary as a standing agenda item was exercising fiduciary oversight, not satisfying an examination requirement. All three were, in fact, doing both simultaneously, but without the mapping that makes that equivalence visible, the regulatory readiness is an accidental byproduct rather than a deliberate outcome.

The Regulatory Readiness Matrix makes that equivalence visible. For each major regulatory category, it identifies what the current environment requires and which governance instruments from this book constitute documented evidence of preparation. For most organizations that have completed the 90-day sprint, the matrix will reveal that they are further along on regulatory readiness than they believed, not because they prepared for regulation, but because responsible governance and regulatory compliance are, in most areas, the same work.

THE REGULATORY READINESS MATRIX: SEVEN CATEGORIES, ONE HONEST ASSESSMENT

The matrix organizes the AI regulatory landscape into seven categories that represent the dominant themes across all major current and pending AI regulatory frameworks, the EU AI Act, the NIST AI Risk Management Framework, federal financial regulatory guidance from the OCC, FDIC, CFPB, and SEC, and the evolving fair lending and employment law applications to AI. Within each category, the matrix identifies the regulatory requirement in plain language, the current enforcement status of that requirement, and, critically, the governance documentation from this book that constitutes evidence of preparation.

Exhibit 6: Regulatory Readiness Matrix

EXHIBIT 6: REGULATORY READINESS MATRIX				
Regulatory Category	**Key Framework / Law**	**Current Status**	**Preparation Level**	**Governance Evidence (What You Need to Show)**
Transparency & Explainability	EU AI Act (Art. 13) NIST AI RMF (MAP 1.5)	EU AI Act: Active NIST RMF: Voluntary	**DOCUMENTED**	AI asset register showing system purpose and data inputs. Board summary showing governance structure. Vendor contracts with transparency obligations.
Human Oversight	EU AI Act (Art. 14) Colorado AI Act (SB 205)	EU AI Act: Active Colorado: Eff. Feb 2026	IN PROGRESS	Override rate documentation for Tier 2/3 systems. Accountability chain showing human oversight responsibilities. Tabletop exercise records.
Non-Discrimination & Fairness	Title VII / ADEA / ADA CFPB Circular 2022-03 EEOC AI Guidance (2022)	Active (US)	IN PROGRESS	Bias audit results for Tier 3 systems in hiring or credit. Output review documentation. Adverse action notice compliance for automated decisions.
Data Governance & Privacy	GDPR (EU) CCPA (California) HIPAA (healthcare)	GDPR: Active (if EU ops) CCPA: Active (CA) HIPAA: Active (if PHI)	**DOCUMENTED**	Data input/output documentation in asset register. Privacy review records. Vendor data processing agreements. DPIA documentation for high-risk AI systems.
Risk Management & Documentation	OCC Bulletin 2021-12 FDIC FIL-22-2022 NIST AI RMF	Active (banking) Voluntary (general)	**DOCUMENTED**	Risk tier classification with documented rationale. TCAIO analysis for material systems. Governance policy formally adopted. Quarterly board summary on file.
Incident Response & Reporting	SEC Cyber Disclosure EU AI Act (Art. 73) State breach laws	SEC: Active (public co) EU AI Act: Active State laws: Active	IN PROGRESS	Written incident response protocol. Tabletop exercise documentation. Incident log in board governance summary. Defined escalation thresholds to board and regulator.
Third-Party Vendor Oversight	OCC Bulletin 2023-17 FDIC FIL-29-2023 EU AI Act (Art. 25)	Active (banking/EU) Best practice (general)	**NOT STARTED**	Five contract non-negotiables in place for Tier 2/3 vendors. Vendor performance review cadence documented. Retraining notification clause in executed contracts.
Update quarterly. Have General Counsel confirm Status column. Update Preparation Level from your actual governance documentation. Add to Section 6 of the Board AI Governance Summary.				

The matrix is not a legal compliance assessment. It is a governance readiness map. Your General Counsel should confirm the status column, regulatory status changes, and the matrix should reflect the current state of enforcement rather than the state at the time this book was written. The preparation level column, however, is yours to assess based on your actual governance documentation. It is the column that converts the matrix from a regulatory tracking tool into an organizational self-assessment instrument.

Regulatory Category	Requirement Summary & Key Sources	Prep Level. Governance Evidence
Transparency & Explainability PENDING	**EU AI Act (Art. 13), NIST AI RMF (GOVERN), CFPB AI Guidance** Requires disclosure that AI is being used in consequential decisions; documentation of system purpose and decision logic available to affected persons.	**DOCUMENTED** *Your AI asset register documents what each system decides and for whom. Column 2 (Business Function) and Column 4 (Decision Authority) are the transparency record.*
Human Oversight ACTIVE	**EU AI Act (Art. 14), FDA AI/ML Guidance, OCC AI Guidance** High-risk AI systems must have meaningful human oversight capability; override mechanisms must exist and be exercised; systems must not disable human intervention	**DOCUMENTED** *Your tier model's override rate monitoring, Tier 2 and Tier 3 override protocols, and accountability chain Link 3 (Business Owner) constitute documented human oversight.*
Non-Discrimination & Fairness ACTIVE	**CFPB, HUD, DOJ AI Fair Lending; EEOC AI Hiring Guidance; EU AI Act (Art. 10)** AI systems used in credit, employment, and housing must not produce discriminatory outputs; regular bias testing required; documentation of testing methodology mandated	**IN PROGRESS** *Your Tier 2 and Tier 3 output review cadence with demographic pattern analysis directly satisfies this requirement. Document the methodology explicitly.*
Data Governance ACTIVE	**GDPR AI processing requirements, CCPA, HIPAA AI provisions, EU AI Act (Art. 10)** AI systems processing personal data must have documented legal basis; data minimization and accuracy requirements apply; individuals have rights regarding automated decisions	**IN PROGRESS** *Your asset register Column 3 (Data Inputs and Outputs) and vendor contract data ownership clause (Non-Negotiable 3) address this. Add legal basis documentation per system.*
Risk Management & Documentation PENDING	**EU AI Act (Art. 9), NIST AI RMF, OCC Model Risk Management (SR 11-7)** Documented risk management process for high-risk AI; ongoing monitoring; board-level risk reporting; incident documentation and response	**DOCUMENTED** *Your TCAIO model, tier classification documentation, incident log in the governance summary, and accountability chain constitute the risk management documentation this standard requires.*
Board & Executive Accountability ACTIVE	**SEC AI Disclosure Guidance, FDIC/OCC AI Governance Expectations, Delaware fiduciary case law** Board-level visibility into material AI risk; documented executive accountability for AI governance; evidence of informed oversight	**DOCUMENTED** *Your Board AI Governance Summary, Executive Sponsor certification, and committee charter mandate are exactly the board visibility documentation this standard describes.*

Third-Party AI Accountability PENDING	**OCC Third-Party Risk Management (2023), EU AI Act importer obligations, CFPB vendor AI guidance** Organizations using third-party AI must exercise oversight of vendor AI governance; vendor performance accountability; due diligence documentation	**DOCUMENTED** *Your vendor contract non-negotiables (Chapter Twelve) and Link 1 (Vendor Accountability) in the accountability chain directly satisfy this requirement.*

Update frequency: *Review and update this matrix quarterly as part of the Board AI Governance Summary preparation. Have your General Counsel confirm the status column; update the prep level column from your governance documentation. The matrix is Section 6 of the governance summary.*

The most important column is the third one: the governance evidence. If your governance program has produced the documentation listed, you have regulatory evidence. If it has not, the gap is not a regulatory gap, it is a governance gap that also happens to be a regulatory gap. Closing it improves your governance and your regulatory posture simultaneously, which is what proactive regulatory readiness looks like in practice.

TRANSPARENCY & EXPLAINABILITY

WHY REGULATORS CARE

Regulators require that organizations be able to explain what AI systems are doing, to whom, and why, in terms that affected individuals and examiners can understand. This requirement is principle-based rather than rule-based in most jurisdictions, which means the standard is sufficiency of explanation rather than compliance with a specific documentation format.

HOW YOUR GOVERNANCE WORK ALREADY ADDRESSES IT

Your AI asset register's Column 2 (Business Function) and Column 4 (Decision Authority) document what each system does and the level of autonomy it operates with. Your governance summary Section 1 provides the board-level transparency record. The accountability chain documents who is responsible for each system's outputs.

WHAT MAY STILL NEED DOCUMENTATION

If your organization uses AI in decisions that directly affect individuals, credit, employment, benefits, healthcare, add a one-paragraph plain-language explanation of each system's decision logic to the asset register. This explanation does not need to be technical. It needs to be understandable to the affected person and to an examiner who is not an AI specialist.

HUMAN OVERSIGHT

WHY REGULATORS CARE

The human oversight requirement is the most consistently enforced theme across all major AI regulatory frameworks: AI systems making or heavily influencing consequential decisions must have meaningful human oversight, and that oversight must be more than nominal. An override mechanism that is never used, or a human review step that is completed in thirty seconds without genuine engagement, does not satisfy the requirement.

HOW YOUR GOVERNANCE WORK ALREADY ADDRESSES IT

Your tier model's override rate monitoring is the most direct governance response to the human oversight requirement. The operational reality test from Chapter Five, which reclassifies systems whose override rates have fallen below the reclassification threshold, is exactly the kind of active oversight monitoring that regulators are looking for. Your accountability chain's Link 3 (Business Owner) documents who is responsible for ensuring human oversight is genuine.

WHAT MAY STILL NEED DOCUMENTATION

Document the override protocol for each Tier 2 and Tier 3 system explicitly: what the override process requires, who has override authority, and how override decisions are recorded. The documentation should be specific enough that an examiner could verify that the protocol is followed by reviewing a sample of output reviews.

NON-DISCRIMINATION & FAIRNESS

WHY REGULATORS CARE

Fair lending, fair housing, and equal employment opportunity requirements are applying to AI systems in their respective domains with increasing specificity. The enforcement trend, across CFPB, DOJ, HUD, and EEOC, is toward requiring organizations to proactively test AI systems for discriminatory outputs rather than responding reactively to complaints.

HOW YOUR GOVERNANCE WORK ALREADY ADDRESSES IT

Your Tier 2 and Tier 3 output review cadence, which includes demographic pattern analysis and bias assessment as standard review elements, is the primary governance response to this requirement. The incident log in your governance summary documents any findings and their resolution.

WHAT MAY STILL NEED DOCUMENTATION

If your organization uses AI in credit, employment, housing, or insurance decisions, the demographic pattern analysis in your output reviews should be documented with enough specificity to satisfy a fair lending examination: the protected classes analyzed, the methodology used, the findings, and the remediation taken for any adverse findings. General statements that bias testing was conducted are not sufficient. Examination-ready documentation specifies what was tested, how, and what was found.

DATA GOVERNANCE

WHY REGULATORS CARE

AI systems processing personal data create data governance obligations that interact with, and in some cases are more demanding than, the organization's general data governance framework. The legal basis for AI processing of personal data, the data minimization and accuracy requirements that apply to training data, and the rights of individuals regarding automated decisions are all areas of active regulatory development.

HOW YOUR GOVERNANCE WORK ALREADY ADDRESSES IT

Your asset register's Column 3 (Data Inputs and Outputs) documents what data each system processes. Your vendor contract's data ownership clause (Non-Negotiable 3) addresses post-termination data rights. Your data governance and security overhead in the TCAIO model (Category 3) captures the cost of the data governance work the AI program requires.

WHAT MAY STILL NEED DOCUMENTATION

Add a legal basis notation to Column 3 for each system processing personal data: the legal basis under which the organization is processing that data for AI purposes (consent, legitimate interest, legal obligation, or other applicable basis under the relevant privacy framework). For systems subject to individual rights regarding automated decisions, primarily credit, employment, and benefits AI in regulated domains, document the process for responding to individual rights requests.

THE REGULATORY POSTURE SHIFT

An organization that can hand an examiner the Regulatory Readiness Matrix in the first ten minutes of an examination has demonstrated something more valuable than regulatory compliance: it has demonstrated that governance is organizational infrastructure rather than regulatory response. That demonstration affects every subsequent interaction with the examiner. It converts the examination from an investigation of whether you have done the required work into a review of how well you have done work you have clearly done. That conversion is worth building for.

THE ORGANIZATION THAT WAS ALREADY READY

Illustrative Scenario: *The following scenario is representative of documented patterns in executive AI governance. Names, organizations, and specific details are illustrative.*

A financial services organization with approximately $4 billion in assets had been building its AI governance program for sixteen months when its primary federal regulator announced an industry-wide AI governance initiative. The initiative was not an enforcement action. It was a request for information: the regulator would be asking organizations across the industry to describe their AI governance practices, and the responses would inform future supervisory guidance.

The organization's Chief Risk Officer read the announcement on a Tuesday morning. She called the CFO and said: "We've been asked to describe our AI governance practices. I want to show you something." She pulled up the Regulatory Readiness Matrix. She walked through each of the seven categories. For five of them, the preparation level was 'documented', the governance work that constituted regulatory evidence already existed. For two, the demographic pattern analysis documentation and the data processing legal basis notation, the preparation level was 'in progress.'

The CFO looked at the matrix for a moment. Then she said: "How long will it take to close the in-progress items?" The CRO said: "Three weeks for the demographic analysis documentation. Two weeks for the legal basis notations. The work is already done, we just need to formalize the documentation."

The CFO said: "Do it. Then let's submit a response that reflects where we actually are."

The response to the regulator's request for information was seventeen pages. It included the Regulatory Readiness Matrix as an exhibit, with all seven categories at 'documented' status. It included the AI asset register, the governance policy, three consecutive Board AI Governance Summaries, the accountability chain documentation for the organization's four Tier 3 systems, and the vendor contract amendments that had been executed in the prior six months. It was, the CRO noted when she submitted it, the most complete AI governance response package she had assembled, and it had required five weeks of documentation work on top of sixteen months of governance work.

Two months after submission, the organization received a letter from the regulator acknowledging its response and noting that the organization's governance framework represented a comprehensive and well-documented approach to AI risk management consistent with supervisory expectations. The letter noted that the organization's practices in two specific areas, the TCAIO model and the accountability chain structure, would be referenced as examples of effective practice in the upcoming supervisory guidance.

"We didn't build our governance program to be a regulatory best practice. We built it because it was the right way to govern AI."

The regulatory recognition was the result of doing the right work, not of doing regulatory work. That distinction, governance for its own sake, regulatory readiness as its byproduct, is what this chapter's thesis looks like when someone has lived it.

MAPPING YOUR READINESS THIS QUARTER

The Regulatory Readiness Matrix is a quarterly governance instrument, not a one-time assessment. Three actions produce the first version and establish the update cadence.

1. **Populate the matrix for your organization today, column by column.**

Open the matrix template from this chapter. For each of the seven regulatory categories, assess your current preparation level honestly against your actual governance documentation: not what you intend to have, not what you believe exists, but what is documented and findable today. Mark each category 'documented' only if the specific governance evidence cited in the matrix exists in a form you could hand to an examiner in the first ten minutes of an examination. Mark categories 'in progress' if the governance work is underway but not yet in documented form. Mark categories 'absent' if the work has not yet begun. The honest assessment is the governance instrument. A matrix that shows six 'documented' and one 'absent' is more valuable than a matrix that shows seven 'documented' without the evidence to support it.

Most common mistake: assessing preparation level based on the governance work that has been done rather than the documentation that exists. An output review cadence that has been operating for six months but whose methodology and findings have never been committed to writing is not documented governance, it is informal practice. The regulatory standard is documentation, not activity. If the work exists only in people's heads and routine practice, it needs to be documented before it counts as regulatory evidence. Success looks like a completed matrix with honest preparation levels and a gap list that identifies what documentation needs to be formalized before the next examination.

2. **Have your General Counsel confirm the status column and add your industry-specific regulations.**

The matrix in this chapter covers the major cross-industry regulatory categories. Every industry has additional AI regulatory requirements specific to its regulatory environment: banking organizations have OCC and FDIC AI guidance, healthcare organizations have FDA AI/ML oversight frameworks and HIPAA AI provisions, insurance organizations have state insurance commissioner AI guidance, and so on. After completing the initial self-assessment, schedule a sixty-minute session with your General Counsel to confirm that the status designations reflect the current regulatory environment for your industry; add any industry-specific regulatory categories that are material to your AI portfolio; and identify any regulatory development that is likely to require a board decision or governance action in the next two quarters.

The status column is a legal question; confirm it with counsel. The preparation level column is a governance question; own it as the Executive Sponsor. These are distinct assessments requiring distinct expertise. The preparation level column that lives in Legal's files rather than the Executive Sponsor's governance folder is a regulatory readiness assessment that no one is acting on. Success looks like a completed matrix with status confirmed by counsel and preparation levels assessed by the Executive Sponsor, reviewed together in a sixty-minute quarterly session before each Board AI Governance Summary is finalized.

3. **Add the matrix to Section 6 of the Board AI Governance Summary and update it quarterly.**

Section 6 of the Board AI Governance Summary (Regulatory Horizon) was designed in Chapter Eight as a text section describing upcoming regulatory developments. The Regulatory Readiness Matrix replaces that text section with a structured, scannable, board-readable document that gives the committee the same information in a format they can interrogate in real time. At each quarterly board presentation, the updated matrix, with status confirmed by counsel and preparation levels assessed by the Executive Sponsor, is the Section 6 exhibit. The matrix's year-over-year evolution, from more 'absent' and 'in progress' cells to more 'documented' cells, is the visual record of the organization's regulatory readiness trajectory. A board that has watched that progression over four quarters has watched its organization move from regulatory exposure to regulatory leadership.

Both columns need quarterly review, not just the status column. Regulations move from proposed to pending to active on their own schedule. The organization's preparation level moves only when governance work is completed and documented. A matrix where the regulatory status column advances while the preparation level stays flat is documenting increasing exposure, not increasing readiness. The two columns should move together, or the gap between them is your governance priority for the quarter. Success looks like a matrix where no regulation has been active for more than two quarters without a corresponding preparation level of "documented."

THE GOVERNANCE THAT LEADS

The thirty-eight hours are not a compliance budget. They are a governance investment. The nine-hour examination package is not the preparation, it is the evidence that preparation is complete. That distinction is the entire argument of this chapter: the accountable executive does not build governance to satisfy regulation. They build governance because it is the right structure for a high-consequence AI program, and when the regulator arrives, they discover that the structure they built is already the answer.

The CRO from the $4 billion organization did not spend sixteen months building a regulatory response. She built a governance program. The regulatory recognition, the supervisory letter naming her practices as effective, the TCAIO model cited as an example, was not the goal. It was the result. Organizations that lead the regulatory conversation are those that built their governance before the regulation arrived to require it. They did not prepare for regulation. They prepared for governance. The distinction, as it turns out, is everything.

Regulatory readiness is not a legal function. It is a governance product. The organizations that lead the regulatory conversation are those that built their governance program before the regulation arrived to require it, and the accountable executive who has done that work will not find regulation arriving to demand it. They will find regulation arriving to confirm it.

The regulatory landscape rewards exactly one posture: the organization that built governance before regulation required it. Everything in this book, the register, the tier model, the TCAIO analysis, the accountability chain, the board summary, has been building toward that posture. Chapter Fourteen addresses the final question: not whether you built it, but whether you built it in a way that survives you.

> ***The accountable executive does not prepare for regulation. They discover that their governance work already constitutes preparation, and that discovery is the reward for having done the right work first.***

BOARDROOM TAKEAWAY | *Knowing What You Have*

What a board member should be able to ask:

- Can you produce a complete list of every AI system in this organization, with a named owner for each, within seventy-two hours?
- Are there any AI systems in Column 5 of our register that are currently unassigned?

What a CFO or risk leader should have documented:

- A completed AI Asset Register with all five columns populated for every known system
- A cross-functional task force formed and tasked with the full inventory detection process

What evidence should exist:

- The register itself: a dated document showing every system, its decision authority, and its named accountability owner
- At least one Shadow AI system identified and added to the register, demonstrating that the detection passes were completed

CHAPTER FOURTEEN: PART IV: THE HORIZON

The Culture That Governs Itself

Embedding Accountability So It Outlasts Any Single Leader

The new CFO found the governance program in a folder on the shared drive. One folder. One person's work. Two years of it.

> **Illustrative Scenario:** *The following scenario is representative of documented patterns in executive AI governance. Names, organizations, and specific details are illustrative.*

Her predecessor had built something genuinely impressive: a completed AI asset register, tier classifications for every system, TCAIO analyses for the Tier 3 portfolio, accountability chain documentation, five consecutive Board AI Governance Summaries. The program had been recognized at a risk management conference. The prior CFO had been invited to speak about it. When she left for a larger organization, the board expressed regret. The technology committee chair said the AI governance program was among her most significant contributions.

The new CFO reviewed the folder in her first two weeks. She found documents that were fourteen months old, a tier classification for a system that had since been decommissioned, an accountability chain that named three people who had left the organization, and a governance summary template that referenced a board committee structure that had been reorganized four months prior. The program was not merely stale. It was structurally disconnected from the organization it had been built to govern.

She called the Chief Risk Officer, who had been in the role for six months and had received no briefing on AI governance when she started. She called the board risk committee chair, who said he had noticed the governance summaries had stopped arriving but had assumed the new CFO was settling in. She called the technology team, who said they had continued deploying AI systems on the existing roadmap and had not thought to ask whether the governance program was current.

The AI asset register had three new systems that had never been added to it. One of them was Tier 3. It had no named accountable owner. It had been running for seven months.

What the prior CFO had built was a governance program. What the organization had was a governance champion.

The moment the champion departed, the program existed only in a folder, disconnected from the people, processes, and board mechanisms that would have made it self-sustaining. The organization had not lost a document. It had lost the institutional memory that the document was meant to outlast.

GOVERNANCE BUILT ON A PERSON

Every AI governance program I have observed was built by someone. A CFO, a Chief Risk Officer, a General Counsel, a COO with a specific governance commitment. That person provided the organizational will that made the program real: they prioritized it in competition with other demands, they secured the executive alignment that the task force required, they presented the governance summaries with the command that produced board confidence, and they maintained the institutional knowledge that kept the program current as the organization's AI portfolio evolved. Without that person, the program does not maintain itself. It ages.

This is not a failure of carelessness. It is the natural result of building governance programs the same way we build every other organizational initiative: around a leader whose commitment drives the work. That approach is appropriate for projects. It is insufficient for governance, which is not a project. Governance is a permanent organizational function that must survive leadership transitions, reporting structure changes, and the gradual attrition of institutional memory that occurs whenever the people who built something move on.

> ***Governance that depends on a champion is governance that is one departure away from collapse. The organizations that lead on AI accountability have embedded it into their institutional DNA.***

The cost of champion-dependent governance is not merely the governance gap that a departure creates. It is the false confidence that precedes the departure. The organization that has a strong governance champion believes it has a strong governance program. When the champion departs, the board discovers that what it had was a strong governance presenter, someone who knew the program deeply, maintained it personally, and carried its continuity in their own expertise. The program's institutional health was never tested because the champion never left. Until they did.

The four mechanisms in this chapter convert a champion-dependent governance program into an institutionally embedded one. Each mechanism addresses a specific fragility in programs built around individuals: the policy gap (governance that lives in a person's judgment rather than organizational policy), the role gap (accountability that attaches to a person's name rather than a position), the board gap (oversight that runs on a champion's initiative rather than a standing mandate), and the culture gap (governance that is practiced by a few rather than expected of the many). Closing these four gaps does not eliminate the value of governance champions. It ensures that when a champion departs, what they built does not depart with them.

THE GOVERNANCE EMBEDDING FRAMEWORK: FOUR MECHANISMS

The mechanisms build on each other in sequence: policy establishes what is required; role specification determines who carries the requirement into each transition; board mandate makes the oversight obligation independent of any individual's initiative; culture markers confirm that governance has moved from program to expectation. All four must be in place. Three out of four is champion-dependent governance with better documentation.

> ***MECHANISM 1, POLICY CODIFICATION***
>
> ***WHY THIS MECHANISM MATTERS***
>
> *Policy converts individual governance judgment into organizational governance requirement. When the TCAIO model is required by policy for every new AI deployment above a defined threshold, the next CFO does not need to know that the*

prior CFO valued full-cost visibility. The policy requires it. When the tier classification protocol is specified in governance policy, the next Risk Officer does not need to have been trained by the prior Risk Officer to know that classification is mandatory. The policy specifies it. Policy codification is the mechanism that makes governance requirements independent of the individuals who invented them.

HOW TO EMBED IT

The AI governance policy drafted in Month Three of the 90-day sprint should specify, at minimum: the tier classification requirement for every AI deployment; the TCAIO estimate requirement for every deployment above a defined cost or risk threshold; the accountability chain requirement for every Tier 2 and Tier 3 system; the vendor contract standard provisions that are non-negotiable; the incident escalation protocol and the timeframes it requires; and the board reporting cadence and the document standard it requires. Each requirement should be specific enough that a compliance function could audit against it without consulting the person who wrote it.

INSTITUTIONAL MARKER

The policy is formally adopted by board or executive resolution, published in the organization's policy library, assigned a review date, and referenced in the onboarding materials for every role with AI governance obligations. A policy that exists but is not findable is not an institutional mechanism, it is archived governance intent.

MECHANISM 2, ROLE SPECIFICATION

WHY THIS MECHANISM MATTERS

Accountability chains that name people rather than positions collapse when those people leave. The Technical Owner who is named in the register by name leaves the organization, and the next person in that role does not know they hold a governance designation. The Executive Sponsor whose name is on the governance summary retires, and the transition plan does not include a transfer of AI governance accountability because no one knew it was attached to the role. Role specification converts personal accountability into positional accountability: the governance obligation belongs to whoever holds the position, regardless of who that person is.

HOW TO EMBED IT

For every link in every accountability chain, the governance obligation should be specified in the position description of the role that holds the link. The Technical Owner obligation belongs in the job description of the engineering lead role that holds it. The Executive Sponsor obligation belongs in the job description of the CFO, COO, or equivalent role that holds it. Board committee AI governance responsibilities belong in the committee charter, not in the institutional memory of the board members who established the practice. When a position description specifies an AI governance obligation, every person who takes that position inherits the obligation automatically, and can be onboarded to it as part of standard role transition.

INSTITUTIONAL MARKER

A new leader joining the organization can locate, within their first week, the AI governance documentation for every system in their portfolio, the accountability chain roles they hold, the governance obligations those roles carry, and the prior governance summaries that establish the program's history. If that is not currently true, role specification is incomplete.

MECHANISM 3, BOARD MANDATE

WHY THIS MECHANISM MATTERS

Board oversight of AI governance that runs on a champion's initiative is oversight that runs on a schedule the champion controls. When the champion is present, the governance summary arrives quarterly and the committee engages. When the champion is absent, the summary stops arriving, and the committee, whose obligation to receive it lives nowhere in its charter, does not ask for it. Board mandate converts the governance summary from a gift the Executive Sponsor delivers to an obligation the board committee requires. That conversion is the difference between oversight that depends on goodwill and oversight that is structurally required.

HOW TO EMBED IT

The board committee charter, whether that is the risk committee, the audit committee, or a dedicated technology governance committee, should specify AI governance as a standing obligation: the committee shall receive a Board AI Governance Summary at every quarterly meeting; the committee shall review the summary and document its receipt and any questions or findings in the committee minutes; any Tier 3 system deployment requires committee notification before deployment; any material incident, regulatory inquiry, or enforcement action involving an AI system requires notification to the committee chair within the timeframe specified in the governance policy. These obligations do not run on the Executive Sponsor's initiative. They run on the committee charter's requirement.

INSTITUTIONAL MARKER

A board committee member who joins the committee receives an onboarding briefing on AI governance as part of their committee orientation, covering the governance summary format, the tier model, the accountability chain structure, and the committee's specific oversight obligations under the charter. The committee does not depend on the Executive Sponsor to explain what it is looking at. The committee charter and orientation materials explain it.

MECHANISM 4, CULTURE MARKERS

WHY THIS MECHANISM MATTERS

The most durable form of governance is the kind that operates without requiring anyone to enforce it, where the question 'who owns this system's governance?' is asked routinely in technology reviews, where the TCAIO requirement for new deployments is treated as a standard budget step rather than a governance imposition, where AI incidents are disclosed to the board as a matter of course rather than as a difficult conversation someone must decide to initiate. Culture markers are the observable indicators that governance has moved from a program to an expectation, from something the organization does because someone is watching to something the organization does because it is how the organization operates.

HOW TO EMBED IT

Culture markers cannot be mandated directly, they emerge from the consistent operation of the first three mechanisms over time. But they can be observed, measured, and accelerated. Observe: does governance language appear in technology reviews without prompting? Does the question of accountability chain coverage come up in deployment approvals before the governance team raises it? Do business line leaders reference their tier classifications when discussing AI investments with the CFO? Accelerate: celebrate and publicly acknowledge the first time a business line leader proactively flags a governance concern before it reaches the risk function. Reference the governance program in the annual report or equivalent public disclosure. Include AI governance in executive leadership development and onboarding. The culture markers are not the goal, they are the evidence that the other three mechanisms have taken hold.

INSTITUTIONAL MARKER

When a mid-level manager in a business line, asked to approve a new AI tool for their team, asks 'have we done the TCAIO and what tier will this be?' without being prompted by the governance function, the culture has been embedded. When that question is a source of mild institutional pride rather than administrative friction, the culture has taken hold.

GOVERNANCE EMBEDDING CHECKLIST

12 institutional markers. An embedded governance program exhibits 9 or more.

Governance Marker	Mechanism	Status
New AI deployments require a TCAIO estimate before budget approval	*Policy*	☐ Check
AI tier classification appears in board risk reports as a standing item	*Board Mandate*	☐ Check
Accountability chain roles are defined in position descriptions, not in individual assignments	*Role Specification*	☐ Check
The phrase 'who owns this system's governance?' is asked routinely in technology reviews	*Culture*	☐ Check
Vendor contract standard includes three AI governance provisions as non-negotiables	*Policy*	☐ Check
The Board AI Governance Summary has been presented for three or more consecutive quarters	*Board Mandate*	☐ Check
An incoming executive can locate the AI governance documentation for their portfolio within their first week	*Role Specification*	☐ Check
AI governance is referenced in the organization's annual report or equivalent public disclosure	*Culture*	☐ Check
The Risk Officer role has explicit AI governance responsibilities in the job description	*Role Specification*	☐ Check
AI incidents are reported to the board committee, not only resolved internally	*Board Mandate*	☐ Check
New employees in governance-adjacent roles receive AI accountability orientation within 30 days	*Culture*	☐ Check
The TCAIO model is updated at least annually for every Tier 2 and Tier 3 system	*Policy*	☐ Check

Score: *9–12 = embedded governance program* | *5–8 = governance in progress* | *0–4 = champion-dependent governance*

> ***THE TRANSITION TEST***
> *The most direct test of whether your governance program is embedded: could your successor govern your AI portfolio in their first thirty days, using only the institutional documentation that exists today, without asking anyone who worked with you? If the answer is yes, the program is embedded. If the answer is no, identify the first gap that would produce a failure and close it before it is tested by an actual transition.*

THE PROGRAM THAT SURVIVED

Four years after completing her organization's 90-day governance sprint, the COO who had built it accepted the CEO role at a larger organization. Her departure was planned twelve months in advance. In the final quarter before her transition, she did something that I have seen very few governance champions do: she audited her own program for institutional dependency.

She ran the transition test on herself. She asked: if I left tomorrow, without a transition conversation, without a briefing, without a handover document, what would survive and what would not? The answer was specific and sobering. The AI asset register would survive because it was in the policy library and required quarterly updates as a matter of policy. The TCAIO model would survive because it was required for all new deployments by the same policy. The board governance summary would survive because it was in the committee charter as a standing requirement.

What would not survive: the Risk Officer's understanding of why the tier classification for one specific system had been set at Tier 2 rather than Tier 3, a judgment call she had made and documented in her head but not in writing. The nuanced vendor relationship with the organization's most consequential Tier 3 system's provider, which she had managed personally and which existed in no documented governance record. The board committee chair's trust in the governance program, which had been built through seven consecutive quarters of personal presentations and would need to be rebuilt by whoever followed her.

She spent three months closing those gaps before her departure. The undocumented tier classification rationale was written into the governance record for that system, with the two-dimension assessment explicitly stated. The vendor relationship was formalized: she convened a governance transition meeting with the vendor's account team and her incoming successor, documented the relationship's history, and had the vendor confirm in writing the governance provisions that had been in place informally. She introduced her successor to the board committee chair and presented three months of governance summaries jointly with her, the outgoing Executive Sponsor presenting alongside the incoming one, explicitly transferring the governance narrative.

When she left, the governance program continued. Her successor presented the first independent Board AI Governance Summary six weeks after the transition. The committee chair said afterward that the summary was indistinguishable in quality from those produced under the prior COO, not because the successor was as experienced, but because the governance program had been designed to produce that quality regardless of who was holding the pen.

She said this:

"Every other program I built was dependent on me knowing why it worked. This one was dependent on the policy, the charter, the role specifications, and the documentation. I had made myself removable from it. That is the only kind of governance that counts."

EMBEDDING BEFORE THE CHAMPION LEAVES

The governance program you have built through this book is, at this moment, likely still champion-dependent. It was built by you, it runs on your knowledge, and it will require deliberate institutional embedding to survive your absence. Three actions begin that embedding.

1. **Run the transition test on your current governance program today.**

Answer the four questions the transition test requires: Would a new CFO or CRO be able to locate and understand the AI governance program from the institutional documentation alone, without a briefing from you? Are the accountability chain roles specified in position descriptions, or do they exist only in the asset register as named individuals? Is the Board AI Governance Summary a standing obligation in the committee charter, or does it arrive because you send it? Is the TCAIO requirement specified in governance policy, or does it exist because you require it? The gaps the transition test surfaces are the embedding priorities. Address the ones that would most immediately collapse the governance program in your absence.

Run the transition test now, while you are present and able to close the gaps without pressure. The leader who runs the transition test at the beginning of a succession process discovers that closing institutional gaps while managing a transition is significantly harder than closing them from a position of stability. The test takes two hours. The gap closure it triggers takes weeks. Both are easier before succession is imminent. Success looks like a written list of the three governance gaps most likely to produce a program failure in a leadership transition, each with a named owner and a deadline for closure.

2. **Add AI governance obligations to two position descriptions this month.**

Identify the two roles in your organization whose AI governance obligations are currently most dependent on the individual in the role rather than the position specification: likely the role holding Link 4 (Risk Officer) in the accountability chain, and the role holding Link 5 (Executive Sponsor). Work with HR or Legal to add explicit AI governance language to each position description: the specific systems the role holds accountability for, the governance obligations that accountability carries, and the institutional documentation the role is expected to maintain. When the next person takes that role, the AI governance obligation is part of what they are accepting.

Most common mistake: adding generic AI governance language to position descriptions rather than specific obligations tied to specific systems. 'Responsible for supporting the organization's AI governance program' is not a governance obligation. 'Serves as Executive Sponsor for LoanSense AI and DemandIQ, with accountability obligations as specified in the AI Governance Policy section 4.2' is a governance obligation. Specificity is what makes position description governance language functional rather than decorative. Success looks like two position descriptions, reviewed and approved by HR, that contain AI governance obligations specific enough that the role holder could not plausibly be unaware of them.

3. **Add AI governance to the board committee charter before the next committee meeting.**

Pull the current charter of the board committee that receives the AI governance summary. Review it for AI governance language. If the charter does not specify AI governance as a standing obligation, including the quarterly summary requirement, the Tier 3 deployment notification requirement, and the material incident notification requirement, draft the amendment and bring it to the committee chair for review before the next meeting. A committee charter amendment typically requires a full board vote or committee approval depending on organizational bylaws; your General Counsel can advise on the appropriate process. The amendment is a one-time action that permanently institutionalizes the oversight obligation the committee has been exercising on your initiative.

Treat the committee charter amendment as an immediate institutional priority, not a future governance project. The charter is the most durable governance instrument in your organization's structure. A charter obligation survives CFO transitions, committee membership changes, and board composition evolution. An informal practice survives none of those things reliably. The amendment is a one-time investment that protects everything the governance program has built. Success looks like a committee charter that specifies AI governance obligations clearly enough that a new committee member, reading it during onboarding, understands what the committee is required to do and how often, without a briefing from the person who built the program.

THE ONLY KIND OF GOVERNANCE THAT COUNTS

"I had made myself removable from it. That is the only kind of governance that counts."

The governance champion who built the program in a folder on a shared drive was not a poor governance leader. She was an excellent one. What she had not done was complete the final act of governance leadership: converting the program she had built into an institution that could govern without her. The four mechanisms in this chapter, policy codification, role specification, board mandate, and culture markers, are the instruments of that conversion. Together they answer the question that every governance program must eventually face: not whether it was built well, but whether it was built to last.

The governance champion who built the program in a folder on a shared drive was not a poor governance leader. She was an excellent one. What she had not done was complete the final act of governance leadership: converting the program she had built into an institution that could govern without her. The four mechanisms in this chapter are the instruments of that conversion. Together they answer the question every governance program must eventually face: not whether it was built well, but whether it was built to last.

Governance that depends on a champion is one departure away from collapse. Making yourself removable from the governance program you built is not a diminishment of your leadership. It is its completion. Chapter Fifteen asks what kind of executive you have become by doing this work, and what that means for the organization you will eventually hand over to someone else.

> ***The organizations that lead on AI accountability have embedded it into their institutional DNA. The governance that outlasts you is the only governance that was ever really yours to give.***

BOARDROOM TAKEAWAY | *The Accountable Executive*

What a board member should be able to ask:

- By completing this book, what governance structures now exist in your organization that did not exist before?
- Can you walk into a board meeting and answer the room question without hesitation?

What a CFO or risk leader should have documented:

- A completed FIDUCIAR model self-assessment with honest scoring across all eight obligations
- A written Monday morning action list: the five concrete governance acts to execute within the next five business days

What evidence should exist:

- A board governance record that demonstrates AI governance was built deliberately, not assembled reactively
- A named person accountable for each governance mechanism, ensuring the program outlasts the executive who built it

CHAPTER FIFTEEN: PART IV: THE HORIZON

The Accountable Executive

What You Now Owe. What You Now Own.

> **Illustrative Scenario:** *The following scenario is representative of documented patterns in executive AI governance. Names, organizations, and specific details are illustrative.*

Eighteen months ago, she could not answer the board chair's question.

The question had been simple by board standards: "Can you give us a picture of where we stand on AI governance?" She had known enough to be worried by it. She had not known enough to answer it. She had opened her laptop, pulled up a briefing document that was already six weeks old, and talked for eleven minutes about AI strategy, innovation pipeline, and competitive positioning, none of which was what the board chair had asked for. The committee chair had thanked her and moved to the next item. She had driven home that evening knowing that something was wrong without knowing exactly what, or how to fix it.

Eighteen months later, she sat in the same room, across the same conference table, and answered the same question in four minutes without notes. She named every Tier 3 system. She cited the TCAIO portfolio figure and the actual ROI ratio. She described the one governance finding from the prior quarter and the remediation already underway. She confirmed that the regulatory horizon showed one pending requirement approaching a board decision point and that Legal was engaged. The committee chair asked three questions. She answered all three from memory. The committee approved the governance summary and moved on.

In the car afterward, her General Counsel, who had been in both meetings, said: "You are a different executive on this topic than you were eighteen months ago."

She said: "I know what I'm governing now. That's the difference."

That sentence is the distance this book was written to cover. This chapter does not introduce a new framework. It closes the distance by naming not what you now have but who you now are.

THE GAP BETWEEN KNOWING IT MATTERS AND KNOWING HOW TO DO IT

Understanding that AI governance matters was the prerequisite for everything that followed. But understanding that something matters is categorically different from knowing how to do it, and for most executives, AI governance has lived in the gap between those two states for longer than it should have. The gap is not produced by negligence. It is produced by the absence of a usable architecture: a clear framework, a defined sequence, instruments specific enough to execute and simple enough to govern.

Most AI governance resources address the understanding side of the gap. They explain why AI is risky, why boards should pay attention, why the regulatory environment is changing, why executives who fail to act may face personal liability. All of that is true. None of it tells an executive what to do Monday morning.

The gap between understanding and execution stays open because virtually everything written about AI governance stops at the point where the governance work begins.

> ***You entered this book as someone who understood AI governance mattered. You close it as someone who can govern. That is the distance this book covered.***

The cost of that gap is not abstract. It is the CFO who could not answer the board chair's question. It is the COO who found out about the model failure from a customer complaint. It is the organization that received a regulatory inquiry and discovered that everything it had done to prepare for governance had not yet been committed to institutional form. The gap between understanding and execution is where AI incidents happen, where regulatory findings accumulate, and where the board's confidence in executive leadership on AI quietly erodes. Closing it is not a technical act. It is a governance act, and it is the act that this book was structured to enable.

The question this chapter asks is not whether you have closed the gap. It is what you do now that you have the tools to close it, a question that requires not a framework but an identity decision: the decision to be the kind of executive whose organization can answer the board chair's question, not because someone prepared a presentation but because the governance is real and the executive who built it knows it completely.

THE FIDUCIAR MODEL REVISITED: FROM TOOL TO PHILOSOPHY

The FIDUCIAR model was introduced in this book's architecture as a governance tool: an **eight-element framework for building an AI governance program from nothing. Each element had a specific deliverable, a specific sequence, and a specific governance instrument associated with it. That is how a framework is introduced. It is not how a governance philosophy is lived.**

By the time you reach this page, you no longer use the FIDUCIAR model as a checklist. You use it as a way of thinking about your organization's relationship with AI: a standing orientation toward the governance questions that will continue to present themselves long after the 90-day sprint is finished and the first board summary has been presented. Below is the FIDUCIAR model as a governing philosophy rather than a governance tool: not what it asks you to build, but what it asks you to be.

> ***F, FRAME AS FIDUCIARY OBLIGATION***
>
> ***As a governance tool:*** *Build the case that AI governance is a fiduciary duty, not a technology preference.*
>
> ***As a governing philosophy:***
> *Every time an AI governance question arises, in a budget conversation, a vendor negotiation, a deployment approval, a board discussion, the accountable executive frames it in fiduciary terms before it becomes a technology debate. Not because the framing produces a different answer every time, but because the framing keeps the accountability at the right level. AI governance questions answered at the*

technology level produce technology answers. AI governance questions answered at the fiduciary level produce governance answers. The framing is not rhetorical. It is the organizational habit that keeps the governance program from drifting back into the technology function where accountability is diffuse and board visibility is limited.

I, INVENTORY ALL AI SYSTEMS

As a governance tool: *Build and maintain the AI asset register.*

As a governing philosophy:
The inventory is not a project that ends when the register is complete. It is an organizational posture: the standing assumption that AI systems appear in organizations faster than governance catches them, and that the executive's obligation is to close that gap continuously rather than once. The accountable executive treats the AI asset register the way a CFO treats the balance sheet, as a document that is always in progress, always subject to update, and always the foundational record from which every other governance instrument draws its accuracy. An executive who commissioned a register eighteen months ago and has not reviewed it since does not have a register. They have a historical document.

D, DIAGNOSE RISK POSTURE

As a governance tool: *Apply the tier model and operational reality test.*

As a governing philosophy:
Risk diagnosis is not a classification exercise performed once at deployment. It is a continuous reading of the organization's AI exposure as systems age, override rates shift, deployment scope expands, and the organization's strategic reliance on specific AI systems changes. The accountable executive has internalized the two-dimension tier model to the point where they can perform a preliminary tier assessment in a conversation, not by calculating, but by asking the right two questions: how consequential are the decisions this system affects, and how autonomous is the system in practice? Those two questions, asked routinely in technology reviews and deployment approvals, keep the risk diagnosis current without requiring a formal assessment every time.

U, UNDERSTAND TOTAL COST OF OWNERSHIP

As a governance tool: *Apply the TCAIO model to every material AI deployment.*

As a governing philosophy:
The TCAIO discipline has a second-order effect that is more durable than any individual cost analysis: it changes the conversation that happens before AI deployments are approved. An organization where the TCAIO model is a standing

requirement for new AI investments is an organization where every AI deployment proposal is constructed with full-cost transparency from the beginning, where vendors are asked for full-cost estimates before contracts are signed, where infrastructure and governance overhead are budgeted in advance rather than discovered after, and where the ROI ratio in the business case reflects actual economics rather than subscription-cost economics. The TCAIO model is not just a measurement tool. It is the governance standard that changes what gets proposed, not just what gets approved.

C, CREATE ACCOUNTABILITY CHAIN

As a governance tool: *Populate the six-link chain for every Tier 2 and Tier 3 system.*

As a governing philosophy:
The accountability chain, once built and embedded in position descriptions and board committee charters, changes the organizational culture around AI ownership in ways that no policy statement can. When every person who takes a role with an accountability chain designation knows that the designation comes with the role, that it is not a personal appointment but a positional obligation, the chain functions as organizational infrastructure rather than individual commitment. The accountable executive's long-term goal is not a set of named individuals in a register. It is a culture in which the question 'who owns this system's governance?' has an immediate, role-based answer that does not depend on the executive being in the room to provide it.

I, IMPLEMENT THE 90-DAY AUDIT

As a governance tool: *Execute the sprint and establish the quarterly cadence.*

As a governing philosophy:
The 90-day sprint was the beginning. The quarterly cadence is the governance. An executive who completed the sprint and has not maintained the quarterly board governance summary, the annual policy review, and the ongoing tier classification update has built a governance foundation and then declined to maintain the structure on top of it. The FIDUCIAR model's second I is not the sprint. It is the institutional commitment that makes the sprint's outputs durable. That commitment is measured not by whether the governance summary was presented last quarter but by whether it will be presented next quarter and the quarter after that, regardless of what else is on the executive's agenda.

A, AUDIT AND VERIFY ONGOING

As a governance tool: *Maintain oversight of override rates, vendor performance, and governance policy compliance.*

As a governing philosophy:
The ongoing audit is the governance practice that converts a governance program from a document into a living oversight function. It is the override rate review that reclassifies a system before a regulator does. It is the vendor contract audit that catches a missing retraining notification clause before a model is retrained without notice. It is the board governance summary preparation that reveals an accountability chain gap before an incident activates it. The accountable executive does not wait for the quarterly board meeting to know the state of the governance program. They know it because the audit function is current, because the governance instruments are maintained, and because the program is designed to surface problems rather than conceal them.

R, REPORT TO THE BOARD IN GOVERNANCE LANGUAGE

As a governance tool: *Present the Board AI Governance Summary as a fiduciary document.*

As a governing philosophy:
The board reporting discipline is the governance program's public accountability moment. It is the moment at which everything the organization has built is translated into the language that the board uses to exercise oversight, and the quality of that translation is the measure of the executive's governance command. The accountable executive who can deliver the five statements without notes, answer the board's ten most likely questions from memory, and present a governance summary that the committee chair describes as 'the information we need to exercise meaningful oversight' has completed the FIDUCIAR cycle. Not once. Every quarter. That is what R means at the philosophy level: not a report, but a standing accountability to the board that the AI governance program is real, current, and under executive command.

THE DISTANCE THE MODEL COVERS
The executive who has internalized the FIDUCIAR model as a philosophy rather than a checklist does not ask 'what does the governance program require of me?' They ask 'what does my governance obligation require of the organization?' That inversion, from compliance posture to fiduciary posture, is the identity shift this book was written to produce.
It is the difference between an executive who has a governance program and an executive who governs.

THE EXECUTIVE WHO BUILT IT

I want to return one final time to the CFO whose story opened this book's Introduction, the one who sat in a discovery deposition and could not produce a document in response to the judge's question. What she did afterward is the point of every chapter that came between that deposition and this one.

She built the governance program. Not reactively, not minimally, not in the form of documentation assembled to satisfy a regulatory requirement. She built it as a governance act: deliberately, rigorously, with board-level transparency and institutional permanence. She completed a version of the 90-day sprint in a hundred and twelve days because her organization's complexity required it. She populated the AI asset register herself for the first two weeks, alongside the task force, because she believed the Executive Sponsor who does not understand the inventory cannot govern it. She presented the first Board AI Governance Summary personally, in a meeting where she disclosed three findings that were unflattering to the organization's prior governance posture. She told the board what had been missing, what she had built, and what she would bring them quarterly going forward.

The board chair's response to that first presentation was not the one she had rehearsed for. She had prepared for skepticism. The board chair said: "This is the most honest governance account I have received from a member of this executive team in my eight years on this board. I don't say that as a criticism of what came before. I say it because I want you to know that this is what we've needed and I'm grateful you built it."

She sat in her car for ten minutes afterward without starting it. What she felt in the moment she delivered the summary was what held her there: not anxiety, not the vertigo of a person who has the information but not the words, not the familiar discomfort of technical territory she was only partially confident in. She had felt, she said, like a CFO.

She had felt like the executive whose board could trust her with AI decisions because she knew what she was governing and could account for it completely.

"I had been a CFO for nine years before that meeting," she said. "I had governed finance, I had governed legal risk, I had governed M&A and integration and capital structure."

"I had not governed AI. I had monitored it, approved spending on it, received briefings about it, nodded at it in board meetings. I had not governed it. After that meeting I had. That distinction, monitoring versus governing, is the one I wish someone had given me five years earlier."

The distinction she names, between monitoring something and governing it, has an inward parallel. The inner ledger that produces composure rather than performance under pressure is the subject of *The Mental Balance Sheet.*

She has now presented the Board AI Governance Summary for eleven consecutive quarters. The governance program has survived one CFO transition in the technology function, one board committee reorganization, and one regulatory examination. It survived all three because it was embedded: in policy, in position descriptions, in the committee charter, and in an organizational culture where the question of AI accountability has a standing answer rather than an appointed one. She built a governance program. More than that: she built a governance institution.

Accountability, built deliberately and maintained rigorously, does not just protect organizations from regulatory findings and board credibility gaps. It converts an executive who monitored AI into an executive who governs it. That conversion is available to every executive who reads this book and decides that Monday morning is when it begins.

WHAT MONDAY MORNING LOOKS LIKE

This is not a chapter about what you have not yet done. It is a chapter about what you now have the capacity to do. The implementation section of every prior chapter gave you a specific set of actions tied to a specific framework element. This one gives you five actions tied to no framework element but the most fundamental one: the decision to begin, or, if the sprint is already underway, the decision to complete.

MONDAY MORNING

Five actions. Any one of them starts the work.

1. ***Schedule the kickoff meeting.*** *If you have not yet started the 90-day sprint, today is the day the calendar commitment happens. Everything in this book becomes real at the moment a kickoff meeting exists on the calendar.*
2. ***Run the transition test on your current governance program.*** *Ask: if you left tomorrow, what would survive? The gaps that answer surfaces are the embedding priorities. Own them before a departure forces them.*
3. ***Look at your highest-consequence AI system and name the person accountable for its governance.*** *Say the name aloud. If you hesitate, that hesitation is the governance gap.*
4. ***Pull one vendor contract for a Tier 2 or Tier 3 system and read it for the five non-negotiables.*** *If all five are missing, the amendment conversation begins this week, not at renewal.*
5. ***Tell one person in your organization that AI governance is a fiduciary obligation, not a technology problem.*** *Governance cultures are built one conversation at a time. This is the first one.*

The most common mistake at this stage is not choosing the wrong action. It is choosing none of them because none feels sufficient. The executive who finishes this book and decides to wait until the conditions are right, until the quarter is less busy, until the team has more capacity, until the regulatory picture is clearer, until there is time to do it properly, is making the same choice as the COO who spent fourteen months intending to start a governance program. The conditions will not improve. The AI portfolio will continue to grow. The regulatory environment will continue to develop. The board will continue to ask questions that the organization is not yet equipped to answer.

Any one of the five Monday morning actions is sufficient to begin. The kickoff meeting scheduled is a governance commitment made. The transition test run is a governance gap named. The vendor contract pulled is a governance risk identified. The conversation with a colleague about fiduciary obligation is a governance culture seeded. None of these actions completes the governance program. Each one makes the next action possible.

Success at this stage does not look like a completed governance program. It looks like an executive who has taken one irreversible step toward building one, who has done something that the version of themselves who only understood that AI governance mattered could not have done: acted.

WHAT YOU NOW OWN

The CFO in the Introduction could not produce a document in response to a judge's question. The judge's line, "CFOs do not delegate fiduciary responsibility. They exercise it", was the sentence that opened this book because it named the standard that every executive who governs AI must meet. Not the standard of perfection. Not the standard of knowing everything about artificial intelligence, model architecture, training data, or regulatory rulemaking. The standard of governance: knowing what you have, grading the risk, understanding the cost, naming the accountable person, reporting to the board, and building the structure that outlasts any individual.

You now know how to do all six.

You now know how to produce every document that matters. The asset register that names every system, its tier, its data inputs, and its accountable owner. The TCAIO model that shows what the portfolio actually costs. The accountability chain that puts a name on every link of governance obligation. The board summary that makes all of it visible to the people responsible for oversight. The regulatory readiness matrix that shows where the organization stands before the examiner asks. The incident protocol that converts a failure into a governance record. She could produce none of these. You can produce all of them.

The distance between understanding that AI governance matters and actually governing AI is not a knowledge gap. It is a decision gap. The decision is available right now. Open a blank document. Write down every AI system running in your organization that you can name without help. That list is your governance baseline. Every blank is a gap this book just equipped you to close.

The Conclusion that follows does not add to what you now know. It closes the book with the one thing a governance framework cannot provide: the reason that all of this matters beyond the examination rooms and board tables and regulatory inquiries where its value is most visible. That reason is worth stating once, clearly, before this book ends.

> ***You are not required to become a technologist. You are required to govern like a fiduciary. You now know how. The rest is Monday morning.***

CONCLUSION

From Passive Approver to Informed Fiduciary

THE RETURN

She sat in a deposition room and could not produce a document. The judge's words have not changed: "CFOs do not delegate fiduciary responsibility. They exercise it." What has changed is everything that follows from that standard. When this book opened, that sentence was an indictment. A measure of distance between where a capable, experienced executive stood and where the obligation required her to be. She had governed finance for a decade. She had not governed AI. The gap between those two facts was the gap this book was written to close.

You have closed it. Not by accumulating understanding, you had that when you opened the book. You closed it by building the architecture that converts understanding into governance: the asset register that names what you have, the tier model that grades the risk, the TCAIO model that counts the real cost, the accountability chain that names the responsible person, the board summary that makes the governance visible, and the institutional mechanisms that ensure it outlasts you. The CFO in the deposition room did not have those instruments. You do. That is the distance between the first page of this book and this one.

THE TRANSFER

The FIDUCIAR model was introduced as a governance framework. It closes as a personal commitment.

Read what follows not as a description of what accountable executives do. Read it as the statement of who you are choosing to be.

> ***THE FIDUCIARY COMMITMENT***
> *Read this as if you are speaking it. Because you are.*
>
> ***F,*** *I will frame AI governance as a fiduciary obligation, not a technology preference. When AI questions arise in my organization, I will answer them at the level of governance, not engineering.*
> ***I,*** *I will maintain an accurate, current AI asset register. I will not govern what I have not mapped, and I will not allow the map to go stale.*
> ***D,*** *I will diagnose my organization's AI risk posture honestly, based on how our systems operate in practice, not how they were designed to operate on paper.*
> ***U,*** *I will understand and report the true cost of every material AI deployment. I will not approve AI on a partial cost basis, and I will not allow my board to do so.*
> ***C,*** *I will create a complete accountability chain for every high-consequence AI system, from the vendor contract to the board committee, and I will ensure that chain is populated with named, informed, willing owners.*

> ***I,*** *I will implement the governance structures this work requires and maintain them through a quarterly cadence that does not bend to the pressure of other priorities.*
> ***A,*** *I will audit the governance program continuously, through override rate monitoring, output review, vendor audits, and tabletop exercises, so that the governance is real rather than documented.*
> ***R,*** *I will report to my board in governance language, quarterly, without notes, with the specificity that demonstrates command rather than familiarity. I will be the executive my board trusts with AI decisions.*

THE DECLARATION

You are now the executive in the room who can answer the question. Not the executive who opens a laptop, pulls up a six-week-old briefing, and discusses AI strategy when the board chair asked about AI governance. Not the executive who reaches for notes when the committee asks about the organization's risk posture or accountability structure. You are the executive who names the Tier 3 systems, cites the ROI ratio, describes the accountability chain, and confirms the regulatory horizon, in four minutes, without notes, with the specificity that tells the board that the governance is real and that the executive who built it knows it completely.

You are the executive whose organization can survive an incident, satisfy an examination, and outlast a leadership transition because the governance is embedded in policy, role, board mandate, and culture rather than in any single person's judgment. That executive is not a future aspiration.

That executive is the one who has read this book and decided to build it. The one who is now deciding when to begin.

THE NEXT ACTION

YOUR NEXT ACTION
Schedule your Day One inventory meeting.
Open your calendar. Block ninety minutes in the next ten business days. Label it: AI Governance Kickoff. Invite the five people who need to be in the room. Send it before you close this book. That action, calendar block sent, is the moment the governance program begins. Not when you have more time. Not when conditions improve. Now.

> ***The most dangerous executive in the age of AI is not the one who moves too fast, it's the one who is still waiting for permission to lead.***

What you have built for your organization, you will eventually need to build for yourself, and for the people who report to you. *Indispensable* applies the same discipline at the individual level: what keeps a leader valuable in an AI augmented workplace, and how to stay on the right side of that line. And every governance decision this book taught you to make is itself one entry in a wider ledger of consequential choices a senior leader carries, which is the work taken up in *The Decision Balance Sheet.*

> **Governance Toolkit: Download live, editable versions of every template in this book:**
> brianzahn.gumroad.com/l/jgnqm

Acknowledgments

Thirty years of professional life does not produce a book. It produces the perspective, the pattern recognition, and the hard-won judgment that a book can only attempt to capture.

I am grateful for every organization that trusted me with responsibility before I had fully earned it, every board that asked the difficult questions, every budget cycle that forced clarity, and every industry transition that required me to rebuild my understanding from the ground up. Biotech. Medical devices. Aerospace. Healthcare. Public infrastructure. Each one added something the previous one could not have.

I am grateful for the colleagues who disagreed with me in the right rooms at the right times, for the mentors who modeled what accountable leadership looks like before that phrase became a governance framework, and for the peers who were willing to say out loud what most executives only think privately. Those conversations are in every chapter of this book, even when no names are attached to them.

I am grateful for the challenges that did not go the way I intended. The projects that exposed gaps. The decisions that required revisiting. The moments where the governance structure either held or did not. A career without those moments produces confidence. A career with them produces wisdom. I have tried to offer the latter here.

And I am grateful to everyone who picks up this book with the intention of building something better in their organization. That decision, made before the incident, before the examination, and before the obligation becomes unavoidable, is the one this book was written to support.

This book belongs to every professional who shaped the thirty years that made it possible.

About the Author

Brian H. Zahn is a Chief Financial Officer in Ventura County, California, where he oversees finance, accounting, financial reporting, internal controls, bond financing, and board governance. He brings a multi-industry career spanning biotech, medical devices, aerospace, healthcare, and government special districts.

Over a career at the intersection of governance, financial oversight, and institutional accountability, he has developed a reputation for translating complex regulatory and operational challenges into clear executive frameworks. His multi-industry career has shaped his perspective on how AI governance must be treated: not as a technology problem, but as a fiduciary obligation.

brian.zahn63@gmail.com

Connect at LinkedIn: www.linkedin.com/in/bzahn

Also from Pacific Meridian Press

The titles below are forthcoming. To be notified as each becomes available, visit pacificmeridianpress.com.

The other titles from Pacific Meridian Press take the same approach, structured thinking applied to the decisions that matter most, into other arenas of professional and personal life.

Indispensable: How to Use AI Without Losing What Makes You Valuable

For the managers, operators, and rising leaders who use AI every day and want to stay valuable as it reshapes the workplace around them. The book is the individual companion to this one: the VATE framework for staying on the side of the line where judgment, trust, and execution still come from the person doing the work.

The Decision Balance Sheet: A Framework for Decisions That Deserve More Than Your Gut

For every leader who has had to make a consequential decision with less structure than they would apply to approving a purchase order. The book applies balance sheet discipline, assets, liabilities, and a defensible record, to the high-stakes decisions a career is eventually judged by.

The Mental Balance Sheet: The Inner Ledger That Drives Every Leader's Performance

For the leader who has built the external structures and is ready to apply the same discipline to the inner state behind every executive decision. The book maps the psychological assets and liabilities that determine how an executive shows up under pressure, with the balance sheet metaphor turned inward.

Love Is Not a Care Plan: A Tactical Framework for the Hardest Family Project You Will Ever Manage

For the executive whose parents are aging and who has applied more structure to a quarterly board summary than to the most consequential family project of the next decade. The book brings the same discipline used here, document stack, financial snapshot, threshold conversations, to turning the worry into action.

Appendix

The following frameworks and tools referenced throughout this book are collected here for quick reference and organizational use.

A.1 The FIDUCIAR Model. Summary Reference

The FIDUCIAR model is the governance framework at the center of this book. It translates eight fiduciary obligations into the specific governance actions every executive must take to govern AI responsibly. Each element operates both as a practical tool and as a governing philosophy.

Element	Full Name	Governance Tool	Governing Philosophy
F	**Frame as Fiduciary Obligation**	Build the case that AI governance is a fiduciary duty, not a technology preference.	*Every time an AI governance question arises, the accountable executive frames it in the language of fiduciary obligation, not technical capability.*
I	**Inventory All AI Systems**	Build and maintain the AI asset register across all five columns.	*The inventory is not a project that ends. It is an organizational posture: the standing assumption that AI systems appear faster than governance structures follow.*
D	**Diagnose Risk Posture**	Apply the tier model and operational reality test to every system.	*Risk diagnosis is continuous. As systems age, override rates shift, and new deployments arrive, the tier assignment must reflect actual operational behavior, not designed intent.*
U	**Understand Total Cost of Ownership**	Apply the TCAIO model to every material AI deployment.	*The TCAIO discipline changes the conversation before deployments are approved. It converts AI investment decisions from capability arguments into full financial accountability.*
C	**Create Accountability Chain**	Populate the six-link chain for every Tier 2 and Tier 3 system.	*The accountability chain, embedded in position descriptions and board committee charters, changes organizational culture around AI ownership in ways no policy document alone achieves.*

I	**Implement the 90-Day Audit**	Execute the sprint and establish the quarterly governance cadence.	*The 90-day sprint was the beginning. The quarterly cadence is the governance. An executive who completed the sprint but has not maintained the cadence has built a governance foundation, not a governance program.*
A	**Audit and Verify Ongoing**	Maintain oversight of override rates, vendor performance, and governance policy compliance.	*The ongoing audit converts a governance program from a document into a living oversight function. Override rate reviews that reclassify systems before incidents occur are governance in practice.*
R	**Report to the Board in Governance Language**	Present the Board AI Governance Summary as a fiduciary document, not a technology update.	*The board reporting discipline is the governance program's public accountability moment. It translates everything built internally into the oversight record the board needs to fulfill its own fiduciary obligations.*

The executive who has internalized FIDUCIAR as a philosophy rather than a checklist does not ask 'what does the governance program require of me?' They ask 'what does accountable governance require of this organization, and am I providing it?'

A.2 The FACT AI Governance Score. Self-Assessment

The FACT AI Governance Score is a rapid self-assessment tool that gives executives a working measure of their organization's current AI governance maturity. Score each of the four dimensions below on a scale of 1 to 5. A total score of 16 or higher indicates a functional governance foundation. Below 12 indicates significant exposure.

Dimension	Score 1 (Absent)	Score 3 (Developing)	Score 5 (Established)	Your Score
F. Foundation Do you have a complete, current AI asset register with all systems named, tiered, and assigned to an accountable owner?	No inventory exists. AI systems are known informally.	A partial inventory exists. Some systems are named but not all are tiered or assigned.	A complete register exists, is maintained quarterly, and every system has a named executive owner.	**/5**
A. Accountability Do accountability chains	No formal accountability	Some Tier 3 systems have	All Tier 2 and Tier 3 systems have	**/5**

exist for all Tier 2 and Tier 3 systems, with each link holder confirmed in writing?	assignments exist beyond the IT owner.	named executive sponsors, but chains are not fully documented.	complete six-link chains, documented and confirmed by each holder.	
C. Cost Visibility Have you completed a TCAIO analysis for your material AI deployments covering all seven cost categories?	AI costs are tracked only at the subscription or license level.	Some systems have partial cost models. Governance and insurance costs are not fully attributed.	TCAIO models exist for all Tier 3 systems and highest-consequence Tier 2 systems, with full seven-category analysis.	**/5**
T. Transparency to the Board Does your board receive a Board AI Governance Summary as a standing quarterly agenda item?	The board receives no dedicated AI governance reporting. AI is covered informally in technology updates.	The board receives periodic AI updates, but they are capability-focused rather than governance-focused.	The Board AI Governance Summary is a standing quarterly agenda item, covering all six sections in governance language.	**/5**
TOTAL SCORE	*4 or below: Significant exposure. Begin with A.3 this quarter.*	*8-11: Developing. Prioritize the weakest dimension first.*	*16-20: Established. Maintain the cadence and expand to new deployments.*	**/20**

Use this score as a starting position, not a performance metric. A score of 8 today with a committed 90-day sprint is a better governance posture than a score of 14 with no improvement plan.

A.3 The 90-Day Governance Sprint Checklist

This checklist is extracted from Chapter Nine. Use it as a working document throughout the sprint. Check off each item as it is completed. Do not advance to Month Two until Month One checkpoint is signed off by the Executive Sponsor.

PRE-SPRINT: THE 10 KICKOFF QUESTIONS

Ask all ten questions at the kickoff meeting before the sprint begins. Score: 7 or more answered confidently = strong governance foundation. 4 or fewer = the gaps define your Month One workplan.

Q#	Question	Can Answer Confidently?
1	Can anyone in this room name every AI system currently running in our organization?	Yes / No / Partial
2	Do we have a document that classifies each system by its level of decision autonomy?	Yes / No / Partial
3	For any system that makes decisions affecting customers, employees, or financial outcomes, who is the named executive accountable for its governance?	Yes / No / Partial
4	What is our current total spend on AI across all budget categories including infrastructure, legal, compliance, and governance?	Yes / No / Partial
5	Has our board ever received a document showing our AI risk posture, incident history, and accountability structure on a single page?	Yes / No / Partial
6	If an AI system produced a harmful output today, who would be notified first and within what timeframe would the board be informed?	Yes / No / Partial
7	Do our vendor contracts require the vendor to notify us when the model is retrained or a material performance issue is detected?	Yes / No / Partial
8	Have we assessed whether any AI system operating as advisory has drifted toward autonomous in actual practice?	Yes / No / Partial
9	What regulatory requirements within 18 months apply to our AI portfolio and where are we in preparing for them?	Yes / No / Partial
10	If a regulator asked for our AI governance documentation today, what would we hand them and how long would it take to assemble it?	Yes / No / Partial

MONTH 1 (Days 1-30): INVENTORY AND CLASSIFICATION

Milestone: A complete AI asset register. Every system named. Every tier assigned. Every shadow AI surfaced.

☐	Week 1	Task
☐		Appoint four-person cross-functional task force (IT, Legal, Procurement, Finance)
☐		Confirm Executive Sponsor in writing
☐		Launch spend analysis detection pass
☐		Launch vendor update log review detection pass
☐		Launch technical staff survey detection pass
☐		Set governance calendar for full 90-day sprint with protected time blocks
☐		Brief General Counsel on sprint scope and timeline
☐	**Week 2**	**Task**
☐		Close all three detection passes
☐		Merge results into single master list
☐		Begin populating AI asset register five-column template
☐		Flag all systems whose accountable owner (Column 5) cannot be named
☐		Deliver preliminary register draft to Executive Sponsor
☐		**EXECUTIVE DECISION: Tier 3 owner confirmations by CEO or equivalent**
☐	**Week 3**	**Task**
☐		Complete all five columns for every system in the register
☐		Conduct classification workshops for systems with contested Column 4 designation
☐		Pull override rates for every Tier 1 and Tier 2 candidate
☐		Document two-dimension tier assessment for each system
☐		Apply operational reality test to all candidate systems

☐	Week 4	Task
☐		Finalize tier assignment for every system
☐		Add Accountability Chain column (seventh column) as placeholder
☐		Executive Sponsor reviews completed register
☐		Escalate any system still missing Column 5 owner to CEO authority
☐		**MONTH 1 CHECKPOINT: Executive Sponsor signs off on register accuracy**

MONTH 2 (Days 31-60): RISK TIERING, COST MODELING, AND CHAIN ASSIGNMENT

Milestone: Every system risk-tiered with documented rationale. TCAIO completed for Tier 3 and highest-cost Tier 2 systems. Accountability chains fully populated.

☐	Week 5	Task
☐		Apply AI Risk Tier Matrix to every system in the register
☐		Verify override rates against system logs (not self-report) for all consequential Tier 1 and Tier 2 systems
☐		Reclassify any system whose operational reality diverges from designed designation
☐		Document two-dimension rationale for every tier assignment
☐	**Week 6**	**Task**
☐		Convene one-hour TCAIO cost attribution meeting with Finance, Legal, and IT for each Tier 3 system
☐		Populate all seven TCAIO categories for each Tier 3 system
☐		Calculate actual ROI ratio for each Tier 3 system against original business case
☐		Flag any system with ROI ratio below 0.5 for Executive Sponsor review
☐	**Week 7**	**Task**
☐		Complete TCAIO analysis for highest-cost and highest-consequence Tier 2 systems
☐		Populate accountability chain (all six links for Tier 3, Links 1-5 for Tier 2, Links 1-3 for Tier 1)

☐		Conduct Link 1 vendor contract audit for all Tier 2 and Tier 3 systems
☐		Identify three contract provisions absent from each vendor contract requiring amendment
☐	**Week 8**	**Task**
☐		Conduct two-hour chain confirmation workshop for each Tier 3 system
☐		Confirm acceptance of governance role obligations by each named Link holder
☐		Schedule tabletop exercise for at least one Tier 3 system before Day 90
☐		Deliver Month Two deliverables to Executive Sponsor: tier register, TCAIO summary, chain documentation
☐		**MONTH 2 CHECKPOINT: Executive Sponsor reviews and approves all Month 2 deliverables**

MONTH 3 (Days 61-90): BOARD REPORT, POLICY ADOPTION, AND CADENCE COMMITMENT

Milestone: Board AI Governance Summary drafted and presented. AI governance policy formally adopted. Quarterly cadence established as a standing organizational commitment.

☐	**Week 9**	**Task**
☐		Draft AI governance policy (four to eight pages, governance language not technology language)
☐		Policy must cover: tier model, TCAIO requirement for new deployments, accountability chain structure, board reporting cadence, incident escalation protocol, vendor contract standards
☐	**Week 10**	**Task**
☐		Draft first Board AI Governance Summary using Chapter Eight template with sprint data
☐		**EXECUTIVE DECISION: Governance policy adopted by formal executive or board resolution**
☐		Have draft governance summary reviewed by one non-technical board member before Week 11
☐	**Week 11**	**Task**

☐		Conduct tabletop exercise for highest-consequence Tier 3 system
☐		Document exercise, participants, and findings in governance record
☐		Incorporate tabletop findings into Section 5 of governance summary
☐		Finalize Board AI Governance Summary based on pre-review feedback
☐	**Week 12**	**Task**
☐		Present Board AI Governance Summary to risk or audit committee as standalone agenda item
☐		Confirm formal adoption of AI governance policy
☐		Establish Board AI Governance Summary as standing quarterly agenda item
☐		Document Day 90 governance record: register, TCAIO summary, chain documentation, adopted policy, first board summary, board receipt confirmation
☐		**SPRINT COMPLETE: Board committee has received and documented receipt of governance summary**

A sprint that produces every governance instrument except the board summary has produced everything except the evidence that governance happened. The board summary is the governance record that makes the rest of the work visible and the accountability real.

> **Did the FACT framework give you a new lens on your AI governance posture? Download the free 12-question self-assessment that frames the maturity levels at brianhzahn.com/cfo-assessment. It takes 20 minutes and produces a defensible baseline you can present to your audit committee.**

Glossary of Key Terms

The following terms are used throughout this book and defined here for executive readers who want clear, non-technical language without sacrificing precision.

AI Governance

The organizational structure of policies, accountability assignments, oversight mechanisms, and reporting processes that ensure AI systems are deployed and operated responsibly and in accordance with fiduciary obligations.

AI Asset Register

A structured inventory of all AI systems in use across an organization, capturing system name, business function, data inputs and outputs, decision authority, and accountable owner.

Autonomous AI (Tier 3)

An AI system in which the AI makes decisions without direct human approval at the point of output. Carries the highest governance and liability exposure.

FIDUCIAR Model

The governance framework introduced in this book: Frame, Inventory, Diagnose, Understand, Create, Implement, Audit, Report. Each pillar maps directly to a fiduciary obligation.

LLM (Large Language Model)

A class of AI systems trained on large volumes of text and capable of generating, summarizing, and reasoning about language. Includes tools such as ChatGPT, Claude, Gemini, and others.

Material AI Risk

An AI-related risk that, if realized, would require disclosure under SEC rules, board notification, or regulatory reporting, or that would significantly impair organizational operations.

Shadow AI

AI tools and systems in active organizational use that have not been formally reviewed, approved, or inventoried by governance or finance leadership.

TCAIO (Total Cost of AI Ownership)

The complete cost model for an AI system, including direct licensing, integration infrastructure, data governance overhead, talent, regulatory compliance, insurance, and opportunity cost.

Exhibits

The following exhibits collect all seven governance templates referenced throughout this book into a single reference section. Each exhibit is designed to be used as a working document. Copy the template, populate it with your organization's actual data, and retain each completed version as a governance record.

> **Governance Toolkit: Download live, editable versions of every template in this book:**
> brianzahn.gumroad.com/l/jgnqm

Exhibit 1: AI Asset Register

Referenced in Chapter Four. The foundational governance document. One row per AI system. Complete all five columns before assigning a tier. Update quarterly. Retain as a board-level governance record.

EXHIBIT 1: AI ASSET REGISTER

Column 1 System Name / Vendor	Column 2 Business Function	Column 3 Data Inputs / Outputs	Column 4 Decision Authority	Column 5 Accountable Owner	Tier (1/2/3)
CreditScore AI / VendorCo	Consumer loan underwriting, supplements underwriter review	In: applicant financials, credit history, employment Out: risk score + recommendation	Assistive: AI drives the recommendation; underwriter approves or overrides	Chief Credit Officer [Name]	2
HR Screening Tool / WorkflowAI	Applicant screening for open positions, filters to shortlist	In: résumés, job descriptions, historical hiring data Out: ranked shortlist; candidates below threshold auto-rejected	Autonomous at screening stage: candidates rejected without human review	SVP Human Resources [Name]	3
Meeting Summarizer / NoteAI	Generates summaries of internal meetings; distributed to attendees	In: meeting audio/transcript Out: summary document; no decision authority	Advisory only: human reviews and distributes summary	IT Operations [Name]	1
Fraud Detection / SecureAI	Real-time transaction monitoring, flags and holds suspicious activity	In: transaction data, account history, behavioral patterns Out: block/hold decision executed automatically	Autonomous: system blocks transaction without human pre-approval	Chief Risk Officer [Name]	3

Instructions: One row per AI system. Complete all five governance columns before assigning tier. Tier is determined by decision autonomy and consequence level, not cost or visibility. Update quarterly. Retain as a governance record.

Exhibit 2: AI Risk Tier Model

Referenced in Chapter Five. The two-dimension classification matrix. Assign every system in the asset register to Tier 1, 2, or 3 based on decision autonomy and consequence level. Classify at the higher tier when dimensions point in different directions.

EXHIBIT 2: AI RISK TIER MODEL

		DECISION AUTONOMY		
		Advisory (Human decides; AI informs)	Assistive (AI drives; human approves)	Autonomous (AI decides; no prior approval)
CONSEQUENCE LEVEL	**Low (Internal / reversible outcomes)**	**TIER 1 Advisory Standard monitoring**	**TIER 1 Advisory Standard monitoring**	**TIER 2 Assistive Enhanced oversight**
	High (External / irreversible / regulated outcomes)	**TIER 1 Advisory Standard monitoring**	**TIER 2 Assistive Enhanced oversight**	**TIER 3 Autonomous Executive oversight required**

TIER	Governance Requirements	Review Cadence	Board Visibility	Override Rate Threshold
1	Named accountable owner. Documented purpose and data inputs. Annual review.	Annual	Portfolio summary only	Monitor; no threshold required
2	Named executive sponsor. Documented accountability chain (Links 1-5). Semi-annual review. Override rate tracked.	Semi-Annual	Named in risk section	Flag if below 15% in consequential domain
3	Named C-suite executive sponsor. Full 6-link chain documented and confirmed. Quarterly board reporting. Tabletop exercise required. Incident protocol in place.	Quarterly	Named individually in every board summary	Escalate if below 5%, reclassify as fully autonomous

Classification rule: When decision autonomy and consequence level point to different tiers, classify at the higher tier. Reclassify based on actual override rates, not designed intent.

Exhibit 3: TCAIO Cost Model

Referenced in Chapter Six. The total cost of AI ownership template. Complete one row per AI system across all seven cost categories. Compare actual ROI against the reported ROI from the original business case.

EXHIBIT 3: TCAIO TOTAL COST OF AI OWNERSHIP MODEL				
#	**Cost Category**	**What to Include**	**System Name: [ ]**	**Annual Cost ($ or estimate)**
1	**Direct Licensing & Subscriptions**	Annual SaaS subscription, per-seat licenses, API usage fees, enterprise license costs. The number in the budget line.	$________	
2	**Integration & Maintenance Infrastructure**	Cloud compute, storage, data pipeline maintenance, IT staff time allocated to this system, monitoring tools.	$________	
3	**Data Governance & Security Overhead**	Data classification, access controls, privacy review, security audits, DPA or GDPR compliance work attributable to this system.	$________	
4	**Talent: Governance, Oversight & Retraining**	FTE hours for governance oversight, output review, model retraining coordination, vendor management. Use fully loaded cost rate.	$________	
5	**Regulatory & Compliance Cost**	Legal review hours, compliance assessments, regulatory filing preparation, external counsel attributable to this system.	$________	
6	**Insurance & Liability Reserve**	Premium allocation for errors & omissions, cyber, and D&O coverage attributable to AI exposure. Pro-rate if portfolio-level.	$________	
7	**Opportunity Cost of Attention**	Executive and board hours spent on governance, oversight, and reporting for this system. Fully loaded rate x hours/year.	$________	
	TOTAL TCAIO		$________	
	Reported ROI (from original business case)		________x	
	Actual ROI = (Annual Benefit - TCAIO) / TCAIO		________x	

Complete one template per AI system. Aggregate to portfolio level for board reporting. If actual ROI is below 1.0x, the system is cash-flow negative. Present to Executive Sponsor before next board cycle.

Exhibit 4: AI Accountability Chain

Referenced in Chapter Seven. The six-link governance specification. Complete Links 1-6 for Tier 3 systems. Links 1-5 for Tier 2. Links 1-3 for Tier 1. Each link holder must confirm their obligations in writing.

EXHIBIT 4: AI ACCOUNTABILITY CHAIN

System Name: ______________________ *Tier:* ___ *Last Reviewed:* __________

Link	Role	Responsibility	Named Individual / Title	Escalation Trigger
1	**Vendor Accountability (Contract Layer)**	Vendor performance obligations, incident notification, model retraining disclosure, regulatory cooperation.	[Vendor contact name] [Account manager / legal] [Contract effective date]	Any output anomaly, model update, or regulatory inquiry involving this system
2	**Technical Owner (Model Layer)**	System configuration, output monitoring, override rate tracking, model version documentation.	[Name, Title] [Department] [Contact]	Override rate crosses threshold; vendor notification received; system output pattern change detected
3	**Business Owner (Decision Layer)**	Business outcome accountability, use case governance, decisions to expand, restrict, or decommission.	[Name, Title] [Department] [Contact]	Material output affecting customers, employees, or financial outcomes; escalation from Technical Owner
4	**Risk Officer (Governance Layer)**	Risk classification maintenance, governance policy compliance, incident log, board reporting input.	[Name, Title] [Contact]	Regulatory inquiry; incident declaration; tier reclassification required
5	**Executive Sponsor (Fiduciary Layer)**	Executive accountability for this system. Board-level reporting. Final decision authority on governance status changes.	[C-Suite Name, Title] [Contact]	Any incident at Tier 3; regulatory examination request; material output causing organizational harm
6	**Board Committee (Oversight Layer)**	Quarterly board visibility. Receives Board AI Governance Summary. Approves governance policy. Final oversight authority.	[Committee name] [Chair name]	Tier 3 system incident; regulatory enforcement action; material governance gap identified

Complete Links 1-6 for all Tier 3 systems. Links 1-5 for Tier 2 systems. Links 1-3 for Tier 1 systems. Each link holder must confirm their obligations in writing during the chain confirmation workshop.

Exhibit 5: Board AI Governance Summary

Referenced in Chapter Eight. The single-page board governance document. Prepare quarterly. Distribute as a standalone agenda item to the risk or audit committee. Retain every version as a permanent governance record.

EXHIBIT 5: BOARD AI GOVERNANCE SUMMARY			
[Organization Name] | Board AI Governance Summary | Quarter Ending [Q/YYYY] | Prepared by: [Executive Sponsor]			
SECTION 1 AI Asset Summary	Total AI systems: 14 Tier 3 (Autonomous): 2 Tier 2 (Assistive): 5 Tier 1 (Advisory): 7 New this quarter: 1	Changes from prior quarter: + 1 new Tier 2 deployment (HR screening) 0 decommissioned 0 reclassified	Tier 3 systems named: 1. Automated credit scoring (CreditScore AI) 2. Real-time fraud detection (SecureAI) Both systems have confirmed accountability chains.
SECTION 2 Risk Posture	Highest-risk system: CreditScore AI (Tier 2) Override rate: 18% Status: Within governance threshold	Monitoring status: All Tier 3 systems: current All Tier 2 systems: current Override rate reviews: completed Q3	Risk flags: HR Screening Tool upgrade planned Q4. Pre-deployment governance review scheduled. No systems currently outside governance threshold.
SECTION 3 Incident Log	Q3 incidents: 1 Severity: Low System: Meeting Summarizer (Tier 1) Nature: Transcript error causing summary inaccuracy	Resolution: Identified Day 2. Corrected summary distributed Day 3. No external impact. Vendor notified. Root cause: audio quality threshold.	Governance action taken: Added audio quality check to Technical Owner review protocol. No tier reclassification required. Documented in governance record.
SECTION 4 Cost Summary (TCAIO)	Total AI portfolio cost (TCAIO): $2.4M annually Reported budget line: $890K Governance overhead: $1.51M	Portfolio ROI (actual): 1.8x (positive) Prior quarter: 1.6x Improvement: Decommissioned 2 cash-negative systems in Q2	Systems with negative actual ROI: 0 Systems under ROI review: 1 (document processing tool, Q4 review scheduled)
SECTION 5 Accountability Status	Accountability chains: Tier 3: 2/2 complete and confirmed Tier 2: 5/5 complete Tier 1: Accountable owners named for all 7	Policy compliance: AI Governance Policy adopted: Q1 20XX Last reviewed: Q2 20XX Next annual review: Q1 20XX+1	Open governance actions: 1. Vendor contract amendment (Link 1, Tier 2 HR tool): in negotiation, target Q4 close 2. Tabletop exercise (Tier 3 SecureAI): scheduled Q4
SECTION 6 Regulatory Horizon	Active regulatory requirements: - ECOA / fair lending: compliant - OCC model risk: compliant - State biometric law: not applicable	Upcoming requirements: - Colorado AI Act: effective Feb 2026 Status: Gap assessment completed, remediation in progress - EU AI Act: not applicable (US-only operations)	Regulatory readiness posture: PREPARED for current requirements. IN PROGRESS on Colorado AI Act. No enforcement actions or regulatory inquiries pending.
This document is a board governance record. Prepare quarterly. Retain per document retention policy. Distribute to board risk or audit committee as a standalone agenda item, not embedded in the technology briefing.			

Exhibit 6: Regulatory Readiness Matrix

Referenced in Chapter Thirteen. The quarterly regulatory assessment tool. Confirm the Status column with General Counsel. Update the Preparation Level column from your actual governance documentation. Add to Section 6 of the Board AI Governance Summary.

EXHIBIT 6: REGULATORY READINESS MATRIX

Regulatory Category	Key Framework / Law	Current Status	Preparation Level	Governance Evidence (What You Need to Show)
Transparency & Explainability	EU AI Act (Art. 13) NIST AI RMF (MAP 1.5)	EU AI Act: Active NIST RMF: Voluntary	DOCUMENTED	AI asset register showing system purpose and data inputs. Board summary showing governance structure. Vendor contracts with transparency obligations.
Human Oversight	EU AI Act (Art. 14) Colorado AI Act (SB 205)	EU AI Act: Active Colorado: Eff. Feb 2026	IN PROGRESS	Override rate documentation for Tier 2/3 systems. Accountability chain showing human oversight responsibilities. Tabletop exercise records.
Non-Discrimination & Fairness	Title VII / ADEA / ADA CFPB Circular 2022-03 EEOC AI Guidance (2022)	Active (US)	IN PROGRESS	Bias audit results for Tier 3 systems in hiring or credit. Output review documentation. Adverse action notice compliance for automated decisions.
Data Governance & Privacy	GDPR (EU) CCPA (California) HIPAA (healthcare)	GDPR: Active (if EU ops) CCPA: Active (CA) HIPAA: Active (if PHI)	DOCUMENTED	Data input/output documentation in asset register. Privacy review records. Vendor data processing agreements. DPIA documentation for high-risk AI systems.
Risk Management & Documentation	OCC Bulletin 2021-12 FDIC FIL-22-2022 NIST AI RMF	Active (banking) Voluntary (general)	DOCUMENTED	Risk tier classification with documented rationale. TCAIO analysis for material systems. Governance policy formally adopted. Quarterly board summary on file.
Incident Response & Reporting	SEC Cyber Disclosure EU AI Act (Art. 73) State breach laws	SEC: Active (public co) EU AI Act: Active State laws: Active	IN PROGRESS	Written incident response protocol. Tabletop exercise documentation. Incident log in board governance summary. Defined escalation thresholds to board and regulator.
Third-Party Vendor Oversight	OCC Bulletin 2023-17 FDIC FIL-29-2023 EU AI Act (Art. 25)	Active (banking/EU) Best practice (general)	NOT STARTED	Five contract non-negotiables in place for Tier 2/3 vendors. Vendor performance review cadence documented. Retraining notification clause in executed contracts.

Update quarterly. Have General Counsel confirm Status column. Update Preparation Level from your actual governance documentation. Add to Section 6 of the Board AI Governance Summary.

Exhibit 7: AI Incident Response Protocol

Referenced in Chapter Eleven. The seven-step incident management framework. Activate on declaration. Document every step as it occurs. Do not close an incident record without Executive Sponsor signature.

EXHIBIT 7: AI INCIDENT RESPONSE PROTOCOL

Activate this protocol when: an AI system produces an output that causes or may cause harm to a customer, employee, or the organization; an output deviates materially from expected behavior; or a regulatory inquiry is received involving an AI system.

Step	Phase	Required Actions	Responsible Link	Documentation Required
1	DECLARE	Activate incident status. Notify Technical Owner and Business Owner. Confirm the incident meets declaration threshold. Time-stamp activation. Do not wait for full picture: declare on reasonable belief.	Technical Owner (Link 2) triggers. Business Owner (Link 3) confirms.	Incident declaration record: system name, time of declaration, who declared, basis for declaration. This record is your Day 1 document.
2	CONTAIN	Identify scope of affected outputs or decisions. Suspend, restrict, or isolate the system if safe to do so. If suspension would create greater harm (e.g., life-safety system), document the decision not to suspend.	Technical Owner (Link 2) Business Owner (Link 3)	Scope document: how many outputs, decisions, or individuals affected. Containment action taken and rationale. If not contained: documented reason.
3	ASSESS	Determine magnitude, affected parties, and potential legal or regulatory implications. Notify Risk Officer. Is this a reportable incident under any contract, regulation, or policy? Engage legal counsel at this step.	Risk Officer (Link 4) Legal Counsel	Assessment memo: magnitude, affected parties, legal and regulatory exposure analysis, recommendation on reporting obligations.
4	COMMUNICATE	Execute notification obligations in order: (a) internal: Executive Sponsor and affected business leadership; (b) vendor: trigger vendor notification clause; (c) external: regulatory or affected party notification per legal advice. Do not communicate externally without legal clearance.	Executive Sponsor (Link 5) Legal Counsel	Communication log: who was notified, when, by whom, in what form. Retain all communications as governance records.
5	INVESTIGATE	Conduct root cause analysis with documentation. Did the failure originate in: model drift, vendor retraining, data input quality, governance gap, or operational error? Preserve all system logs and output records before investigation begins.	Technical Owner (Link 2) Risk Officer (Link 4)	Root cause analysis document: technical findings, contributing governance factors, what governance controls did or did not function as designed.

6	**REMEDIATE**	Restore, retrain, restrict, or decommission the system based on root cause findings. Remediation must include: correction of affected outputs where possible; system fix or replacement; governance policy update if a gap contributed.	Business Owner (Link 3) Technical Owner (Link 2)	Remediation plan and completion record: what was done, when, by whom, and how the fix was verified.
7	**DOCUMENT & CLOSE**	Build the complete incident audit trail before closing the incident record. Present summary to Executive Sponsor for sign-off. Add to Section 3 (Incident Log) of the next Board AI Governance Summary. Do not close the incident record without Executive Sponsor signature.	Executive Sponsor (Link 5) Board Committee (Link 6) for Tier 3	Closed incident record: all steps documented, root cause confirmed, remediation verified, board summary updated, Executive Sponsor signature. Retained per document retention policy.

Declaration threshold (define before an incident occurs): An incident must be declared when an AI system output causes harm, potential harm, or material deviation from its governed operating parameters. Threshold applies regardless of severity. Severity determines escalation speed, not whether to declare.

Notes

Sources are organized by chapter. All illustrative scenarios are composites based on documented patterns in AI governance, regulatory enforcement, and corporate accountability. Specific details are not drawn from any single identified matter unless explicitly cited.

Introduction

1. The opening deposition scenario is a composite illustrating documented patterns in AI-related financial services litigation. For real cases involving AI-based coverage denials and algorithmic harm, see Estate of Gene B. Lokken et al. v. UnitedHealth Group et al., Case No. 0:23-cv-03514 (D. Minn., filed Nov. 14, 2023), alleging UnitedHealth's naviHealth nH Predict algorithm denied elderly Medicare Advantage patients medically necessary care with a documented 90% error rate. For AI hiring discrimination, see Mobley v. Workday, Inc., Case No. 3:23-cv-00770-RFL (N.D. Cal.), in which the court on July 12, 2024 allowed claims to proceed that Workday's AI-powered applicant screening tools discriminated on the basis of race, age, and disability. Mobley v. Workday, Inc., 740 F.Supp.3d 796 (N.D. Cal. 2024).
2. The fiduciary standard that executives may not delegate fiduciary responsibility reflects established corporate governance doctrine. See In re Caremark Int'l Inc. Derivative Litigation, 698 A.2d 959 (Del. Ch. 1996) (foundational board oversight duty for compliance failures); Stone v. Ritter, 911 A.2d 362 (Del. 2006) (extending Caremark to require good faith oversight of corporate operations); Smith v. Van Gorkom, 488 A.2d 858 (Del. 1985) (informed decision-making standard under the Duty of Care). For the principle that fiduciary duties apply to AI governance in the same manner as other operational risks, see Marchand v. Barnhill, 212 A.3d 805 (Del. 2019) (board liability for failing to monitor food safety risk, establishing that Caremark applies to operational compliance, not only financial fraud).
3. Regulatory movement from advisory to mandatory: EU AI Act, Regulation (EU) 2024/1689 of the European Parliament and of the Council, Official Journal of the European Union L 2024/1689, July 12, 2024, effective August 1, 2024 (eur-lex.europa.eu). NIST AI Risk Management

Framework, NIST AI 100-1 (January 2023) (doi.org/10.6028/NIST.AI.100-1). FTC Policy Statement on Artificial Intelligence, November 2023 (ftc.gov). State-level legislation: Colorado Artificial Intelligence Act, SB 24-205, signed May 17, 2024, effective February 1, 2026; Illinois Artificial Intelligence Video Interview Act, 820 ILCS 42, effective January 1, 2020; California AB 2013 (AI training data transparency), signed September 2024.

4. Federal litigation involving AI-related corporate and executive accountability is actively developing. Key cases: (1) Estate of Gene B. Lokken et al. v. UnitedHealth Group et al., Case No. 0:23-cv-03514-JRT-SGE (D. Minn., filed Nov. 14, 2023), alleging systematic AI-driven denial of Medicare Advantage coverage; on February 13, 2025, the court allowed breach of contract and good faith claims to proceed. (2) Mobley v. Workday, Inc., Case No. 3:23-cv-00770-RFL (N.D. Cal.), allowing claims of AI hiring discrimination to proceed under an 'agent' theory of employer liability; on May 16, 2025, the court certified a collective action under the ADEA on behalf of potentially millions of job applicants. (3) Harper v. Sirius XM Radio (2025), a follow-on case alleging AI hiring discrimination under Title VII by an African American applicant rejected from 149 positions. For a comprehensive tracker of AI-related litigation, see the Partnership on AI's AI Incident Database at incidentdatabase.ai.

Chapter One: The Last Executive Standing

5. All scenarios in this chapter, including the eleven-hour deposition and the risk committee chair story, are composites representative of documented AI governance litigation and regulatory examination patterns. See composite label preceding each scenario.
6. Four categories of AI-related executive liability established in US and EU precedent. Category 1 (Discriminatory Outputs): See Mobley v. Workday, Inc., 740 F.Supp.3d 796 (N.D. Cal. 2024), holding that an AI vendor's screening tools may constitute employment discrimination under Title VII, the ADEA, and the ADA when operating as an agent of the employer; EEOC Technical Assistance, 'The Americans with Disabilities Act and the Use of Software, Algorithms, and Artificial Intelligence to Assess Job Applicants and Employees,' May 12, 2022; CFPB Circular 2022-03, Adverse Action Notification Requirements (May 2022). Category 2 (Data Privacy): Irish Data Protection Commission fine of EUR 1.2 billion against Meta, May 22, 2023 (for unlawful EU-to-US data transfers); Netherlands DPA fine of EUR 30.5 million against Clearview AI, 2024 (for GDPR violations involving biometric data); Italy's Garante fine against OpenAI, late 2024 (first GDPR enforcement action against a generative AI company for unlawful data processing). Category 3 (Negligent Deployment): See Estate of Gene B. Lokken et al. v. UnitedHealth Group, Case No. 0:23-cv-03514 (D. Minn.), alleging negligent deployment of an AI system with a documented 90% error rate for denying medically necessary care. Category 4 (Failure of Board Oversight): In re Caremark Int'l Inc. Derivative Litigation, 698 A.2d 959 (Del. Ch. 1996); Marchand v. Barnhill, 212 A.3d 805 (Del. 2019) (extending Caremark to operational risk monitoring).
7. 'Matters Requiring Attention' is a formal supervisory finding category used by OCC, FDIC, and Federal Reserve in safety and soundness examinations. References to MRA findings in this book are illustrative of documented examination outcomes. See OCC Policies and Procedures Manual, Safety and Soundness Examinations.

Chapter Two: Shadow AI

8. The manufacturing company Shadow AI discovery scenario is a composite. The specific revenue figure ($2.4 billion) is illustrative. For research on unmanaged AI tool usage across enterprises, see Stanford AI Index Report 2024 (hai.stanford.edu), documenting enterprise AI adoption patterns and governance gaps; McKinsey Global Survey on AI (2023), noting that over 55% of organizations have adopted AI in at least one business function with governance frameworks lagging deployment. For Shadow AI specifically, see Gartner research on 'ungoverned AI' and enterprise SaaS sprawl, noting that the average enterprise uses hundreds of SaaS applications, many with embedded AI features that have not been formally reviewed.

Chapter Three: The Fiduciary Reframe

9. The CEO/CTO enforcement action scenario and the technology committee scenario are composites representative of documented governance accountability patterns. See composite labels preceding each scenario.
10. Duty of Care, Duty of Loyalty, and Business Judgment Rule as applied to AI governance contexts. Foundational cases: In re Caremark Int'l Inc. Derivative Litigation, 698 A.2d 959 (Del. Ch. 1996); Smith v. Van Gorkom, 488 A.2d 858 (Del. 1985) (Duty of Care and informed decision-making); Revlon, Inc. v. MacAndrews & Forbes Holdings, 506 A.2d 173 (Del. 1986) (Duty of Loyalty); Stone v. Ritter, 911 A.2d 362 (Del. 2006) (good faith and oversight obligations). For application of these duties to technology and AI governance specifically, see Marchand v. Barnhill, 212 A.3d 805 (Del. 2019), in which the Delaware Supreme Court held that a board's failure to monitor food safety risks, an operational matter rather than financial fraud, constituted a Caremark violation. Legal commentators have widely applied this precedent to AI governance obligations. See also SEC guidance on material cybersecurity and technology risk disclosure, which creates parallel disclosure obligations for AI-related material risks.

Chapter Five: Risk That Isn't Abstract

11. The underwriting model demographic pattern scenario is a composite illustrating documented algorithmic lending discrimination risks. Regulatory context: CFPB Circular 2022-03, Adverse Action Notification Requirements in Connection with Credit Decisions Based on Complex Algorithms (May 26, 2022); CFPB, 'Interagency Statement on Model Risk Management for Bank Systems Supporting Bank Secrecy Act/Anti-Money Laundering Compliance,' April 9, 2021; Joint Statement on AI in Financial Services by the CFPB, FDIC, Federal Reserve, FDIC, FHFA, NCUA, and OCC (April 25, 2023), noting that AI systems used in credit decisions must comply with the Equal Credit Opportunity Act and other fair lending laws regardless of whether the model is explainable. For the most directly applicable enforcement context, see CFPB Supervisory Highlights (Issue 29, Summer 2023) discussing AI and credit decision oversight.
12. The clinical AI override rate scenario is a composite illustrating automation bias in clinical decision support. For research on automation bias and physician override rates in AI-assisted clinical settings, see: Goddard K, Roudsari A, Wyatt JC, 'Automation bias: a systematic review of frequency, effect mediators, and mitigators,' Journal of the American Medical Informatics Association, 2012; Parasuraman R, Manzey DH, 'Complacency and Bias in Human Use of Automation: An Attentive Review,' Human Factors, 2010; and for AI-specific clinical decision support override rates, see literature in the Journal of the American Medical Association (JAMA) and New England Journal of Medicine (NEJM) on clinical decision support and physician

behavior. The pattern of overrides declining significantly over time in clinical settings without governance intervention is documented in multiple clinical AI deployment studies.

Chapter Seven: The Chain

13. The benefits administrator eligibility miscategorization scenario and the logistics company routing failure scenario are composites. The 'silent retraining' issue described in the logistics scenario reflects a documented governance risk. See EU AI Act Article 13 (transparency obligations for high-risk AI systems); NIST AI RMF, Govern 1.2 (policies for AI risk documentation and monitoring).

Chapter Eight: The Governance Layer

14. The regional bank examination scenario, including the Matters Requiring Attention finding, is a composite illustrating documented OCC and FDIC examination outcomes for organizations with inadequate AI governance board reporting. The examination narrative reflects supervisory practices documented in OCC Bulletin 2021-12 (Model Risk Management) and FDIC FIL-22-2022.

Chapter Nine: The 90-Day Audit

15. The COO regulatory inquiry scenario is a composite. Regulatory information requests regarding AI governance practices are documented across financial services. See, e.g., New York Department of Financial Services (NYDFS) Circular Letter on AI in Insurance (December 2023), requesting documentation of AI governance practices; Federal Reserve and OCC supervisory communications on model risk management expectations for AI systems (OCC Bulletin 2021-12; SR 11-7). The SEC has issued comment letters to public companies referencing AI governance disclosure obligations under existing securities laws.

Chapter Eleven: When It Fails on Your Watch

16. The assertion that every organization deploying AI will eventually experience an AI incident reflects documented statistical certainty about complex systems in production environments. See AI Incident Database, maintained by the Responsible AI Collaborative, cataloging over 1,200 real-world AI incidents across sectors (incidentdatabase.ai); MIT AI Incident Tracker, part of the MIT AI Risk Initiative, which classifies AIID incidents using the MIT Risk Repository (airisk.mit.edu); NIST AI Risk Management Framework, NIST AI 100-1 (January 2023), Section 2.5 on AI system failures in production. For healthcare-specific AI incident rates, see Estate of Gene B. Lokken et al. v. UnitedHealth Group, Case No. 0:23-cv-03514 (D. Minn., 2023), alleging a 90% error rate in a deployed clinical AI system.

Chapter Twelve: The Vendor Negotiation No One Prepares For

17. The General Counsel vendor contract scenario is a composite illustrating documented regulatory examination challenges involving AI vendor contracts. Regulatory context for AI vendor oversight: OCC Bulletin 2023-17 (Third-Party Risk Management); FDIC FIL-29-2023; Federal Reserve SR 23-4. These guidelines establish obligations for financial institutions regarding third-party AI vendor governance.

Chapter Thirteen: Regulation Is Coming. The Accountable Executive Is Already Ready.

18. EU AI Act: Regulation (EU) 2024/1689 of the European Parliament and of the Council, Official Journal of the European Union L 2024/1689, July 12, 2024. Available: eur-lex.europa.eu/legal-content/EN/TXT/?uri=OJ:L_202401689.

19. NIST AI Risk Management Framework: National Institute of Standards and Technology, Artificial Intelligence Risk Management Framework (AI RMF 1.0), NIST AI 100-1, January 2023. Available: doi.org/10.6028/NIST.AI.100-1.

20. OCC Model Risk Management: OCC Bulletin 2021-12 (March 2021). FDIC: FIL-22-2022 (April 2022). Federal Reserve: SR 11-7, Guidance on Model Risk Management (April 2011). Third-party risk: OCC Bulletin 2023-17; FDIC FIL-29-2023; Federal Reserve SR 23-4 (June 2023).

21. FDA AI/ML Guidance: U.S. Food and Drug Administration, 'Artificial Intelligence and Machine Learning (AI/ML)-Based Software as a Medical Device,' January 12, 2021. Available: fda.gov. HIPAA: 45 CFR Parts 160 and 164.

22. State-level AI legislation: Colorado Artificial Intelligence Act, SB 24-205, signed May 17, 2024, effective February 1, 2026 (first US law to directly regulate high-risk AI in credit, employment, insurance, and public benefits); Illinois Artificial Intelligence Video Interview Act, 820 ILCS 42, effective January 1, 2020 (first US state AI employment law); California AB 2013, signed September 2024 (AI training data transparency requirements); California SB 1047 (vetoed by Governor Newsom, September 2024, but signaling continued state legislative activity); New York City Local Law 144, effective July 5, 2023 (requiring bias audits for automated employment decision tools); Texas HB 149 (AI in consumer finance, effective September 2023). For a comprehensive tracker of state AI legislation, see the National Conference of State Legislatures AI tracking page (ncsl.org) and the Future of Privacy Forum state privacy and AI legislation tracker.

23. The $4 billion financial services organization scenario and the CRO regulatory examination scenario are composites representative of documented supervisory practices. Positive regulatory recognition for strong AI governance programs has been noted in OCC and FDIC supervisory communications. See OCC Bulletin 2021-12 (Model Risk Management, March 2021) and FDIC FIL-22-2022 (Supervisory Guidance on Model Risk Management, April 2022), both of which describe examination expectations and note that organizations with comprehensive, documented governance programs receive more favorable examination treatment. For industry-wide AI governance reviews, see Federal Reserve, OCC, FDIC, CFPB, FHFA, NCUA, and OCC Joint Statement on AI in Financial Services (April 25, 2023).

This Notes section reflects the state of publicly available law and regulatory guidance as of May 2026. Laws, regulations, and cases cited are subject to change. Readers are encouraged to consult current legal counsel for jurisdiction-specific obligations. For the most current AI regulatory developments, see the OECD AI Policy Observatory (oecd.ai), the Future of Privacy Forum (fpf.org), and the IAPP AI Governance Center (iapp.org).

www.ingramcontent.com/pod-product-compliance
Lightning Source LLC
LaVergne TN
LVHW022214200926
848481LV00006B/1962
* 9 7 9 8 9 9 5 5 2 0 8 1 8 *